Introduction to Research Methods

To Maia and Ludmilla Antonia
and their bright futures.

Sara Miller McCune founded SAGE Publishing in 1965 to support the dissemination of usable knowledge and educate a global community. SAGE publishes more than 1000 journals and over 800 new books each year, spanning a wide range of subject areas. Our growing selection of library products includes archives, data, case studies and video. SAGE remains majority owned by our founder and after her lifetime will become owned by a charitable trust that secures the company's continued independence.

Los Angeles | London | New Delhi | Singapore | Washington DC | Melbourne

Introduction to Research Methods

A Hands-On Approach

Bora Pajo
Mercyhurst University

Los Angeles | London | New Delhi
Singapore | Washington DC | Melbourne

FOR INFORMATION:

SAGE Publications, Inc.
2455 Teller Road
Thousand Oaks, California 91320
E-mail: order@sagepub.com

SAGE Publications Ltd.
1 Oliver's Yard
55 City Road
London EC1Y 1SP
United Kingdom

SAGE Publications India Pvt. Ltd.
B 1/I 1 Mohan Cooperative Industrial Area
Mathura Road, New Delhi 110 044
India

SAGE Publications Asia-Pacific Pte. Ltd.
3 Church Street
#10-04 Samsung Hub
Singapore 049483

Acquisitions Editor: Leah Fargotstein
Editorial Assistant: Yvonne McDuffee
Development Editor: Eve Oettinger
eLearning Editor: Laura Kirkhuff
Production Editor: Andrew Olson
Copy Editor: Janet Ford
Typesetter: C&M Digitals (P) Ltd.
Proofreader: Sarah J. Duffy
Indexer: Nancy Fulton
Cover Designer: Karine Hovsepian
Marketing Manager: Shari Countryman

Printed in the United States of America

Library of Congress Cataloging-in-Publication Data

Names: Pajo, Bora, author.

Title: Introduction to research methods : a hands-on approach / Bora Pajo.

Description: Los Angeles : SAGE, 2017. |
Includes bibliographical references and index.

Identifiers: LCCN 2017021397 | ISBN 9781483386959 (pbk. : alk. paper)

Subjects: LCSH: Social sciences—Research—Methodology.

Classification: LCC H62 .P235 2017 | DDC 001.4/2—dc23
LC record available at https://lccn.loc.gov/2017021397

This book is printed on acid-free paper.

17 18 19 20 21 10 9 8 7 6 5 4 3 2 1

BRIEF CONTENTS

DETAILED CONTENTS

PREFACE

So why would you consider reading another book on research methods when there are tons of such books readily available in the market?

Truthfully, I may not know the answer to that question. I may share, however, the experience of teaching research methods every year and having the unpleasant aftertaste that students were barely able to follow all the definitions and nuances of research. Or did not quite see the importance of learning research methods, or even the ones who tried still felt like they were learning something unnatural and unrelated to their futures. What is even worse, halfway through the semester, I started to sympathize with struggling students and began to consider their viewpoint that research methods are irrelevant to their future goals. Year after year, I tried doing things differently in the class, creating videos, using humor, visual aids, you name it, but nothing seemed to work. I needed something that students could do on their own, something they could touch and experience. I needed to get them to walk in my shoes. So I experimented with learning by doing, even if the doing was not scientific initially or did not follow strict methodological rules. If I could get them interested in their own studies, in their own mock research, they would certainly try to learn more and, in the process, understand the importance of research methods. This book is truly an attempt to get students to appreciate research methodologies and their endless potential. In a nutshell, my motivation for taking over such a labor-intensive project was to write a practical step-by-step guide on how to conduct research.

Second, I recognize that research is not intuitive, does not come naturally for most of us, cannot be learned by half listening to the professor in class, and most importantly cannot be learned if it is feared and taught in the abstract. There is a real fear among students and faculty alike when it comes to learning or teaching research methodologies. Most of the books available treat the material with the appropriate seriousness the subject deserves. Although these types of textbooks may be very useful to a seasoned researcher, they read like a dictionary to a student who is looking at the art and science of conducting research for the first time. Alternatively, this text aims at simplifying concepts, illustrates them with examples, and attempts to look at research methods in a light-hearted manner. In other words, this book is taking a shot at introducing concepts in a memorable way for the novice researcher.

Finally, this book has become so personal, living within my head for over 2 years now. I have delved into each chapter, have carefully considered every concept, and have even taken a stab at drawing its illustrations. On that note, I hope you will enjoy reading about conducting scientific research.

ACKNOWLEDGMENTS

This book would not have been possible without the immense support, continuous guidance, and high professionalism of everybody at Sage. I could not wish for a better team. Specifically, I want to thank my editor, Leah Fargotstein, whose support and ideas have been so crucial in every step of the book development. Her enthusiasm, trust in me, and attention to detail have been my solid guidance. I want to give a special thank you to my development editor, Eve Simon Oettinger, whose creative ideas were at the core of organizing my writing and keeping me on track. Eve was able to delicately bring organization to what seemed like a mess of individually written pieces. Thank you for your patience and your amazing guidance. I also want to thank Vicki Knight for making this book a reality and encouraging me to do my best. Her words and enthusiasm were so crucial at the beginning of this project when writing felt very much like a lonely process. This book would not have been possible without the amazing work of Leah Mori, whose editing, precision, and attention to detail gave life to the entire book. I want to thank Yvonne McDuffee, Tori Mirsadjadi, Janet Ford, Jennevieve Fong, Laura Kirkhuff, and everyone at Sage for their support and guidance every step of the way during these last 2 years. I would also like to thank the following reviewers who have been such an enormous help throughout this entire process:

Brenda I. Gill, Alabama State University

Bernadette R. Hadden, City University of New York

William Holland, Georgia State University

Jennifer L. Huck, Carroll University

Edward L. Jackiewicz, California State University Northridge

Wesley L. James, University of Memphis

Daphne John, Oberlin College

Shelley Dean Kilpatrick, Southwest Baptist University

Janet Laible, Lehigh University

Robert W. Lancaster, Kentucky State University

Jason LaTouche, Tarleton State University

David A. Licate, University of Akron

Chuck Lubbers, University of South Dakota

James C. Petersen, University of North Carolina at Greensboro

Pierre Pratley, George Washington University

Katherine K Rose, Texas Woman's University

Beverly Ross, California University of Pennsylvania

Daniel A. Sanchez, Ohio State University

Burt Stillar, University of La Verne

Shalini A. Tendelkar, Tuffs University

Richard Williams, University of Notre Dame

Julie A. Winterich, Guilford College

Robert Wonser, College of the Canyons

Mohammad Zannoun, University of Kentucky

I am deeply grateful to Dr. Charles Sarno, who wrote the qualitative chapter (Chapter 11) of this book, for both agreeing to contribute to this project and offering his qualitative expertise. His warm style of writing, clear explanation of qualitative methods, attention to ethical dilemmas, and engaging style are so much appreciated. Dr. Sarno's addition to this book adds the qualitative heart and thinking that is extremely necessary and valuable to students.

In addition, I want to thank my chair and mentor, Laura Lewis, whose encouragement and support during the writing process of this book was extremely valuable. I want to give special thanks to my colleague and friend Maria Garase, who followed closely every step of the process and always offered helpful suggestions. I want to thank my dean, Randy Clemons, who always sympathized with the long process of writing a book, and encouraged me to do my best. Finally, I thank my entire family who followed the ongoing process of developing this book with great enthusiasm and encouragement, especially my husband who kept me going every step of the way.

THE PURPOSE OF RESEARCH

CHAPTER OUTLINE

WHAT WILL YOU LEARN TO DO?

1. Describe scientific research and its purpose in furthering knowledge
2. Summarize two theories of knowledge: falsifiability and the scientific revolution
3. Compare and contrast qualitative, quantitative, and mixed methods
4. Explain the importance of ethics and objectivity in research

SCIENTIFIC RESEARCH AND ITS PURPOSE

We humans are great knowledge accumulators. We love knowing about everything, and these days it is quite easy to obtain knowledge. I start the day by listening to news on the radio while driving into work. My eyes catch a new billboard on the highway—something about hospitals and children's health. To get to my office, I walk across campus, but along the way, my senses are bombarded with advertisements, posters, and all kinds of information that beg for my attention. Finally, inside my office, I boot up the computer. Preparing for class is accompanied by checking email, scrolling through my department's Facebook page, tweeting about the latest *New York Times* article on children diagnosed with attention deficit hyperactivity disorder (ADHD), and, of course, double-checking my Prezi presentation. My eyes also catch some information about a new diet program, a new research methods book, and a new study on children's health. And how can I refuse the latest video from my 10-year-old niece who is programming robots? I am such a proud auntie! Wait! Wasn't that a picture of my friend's newborn twins? How cute!

This is likely to be a familiar scenario in your life as well. We are accustomed to absorbing vast amounts of information every day. But how do we distinguish accurate from inaccurate information? What communications can we actually trust? You will probably agree that some nonscientific knowledge comes from cultural *tradition*, such as how to roast a turkey on Thanksgiving, the right amount to tip a waiter when eating at a restaurant, or even how to dress as a girl or a boy. So **traditional knowledge** is a form of knowledge that we inherit from the culture we grew up in. This includes everything we were taught as children that has become part of who we are and how we behave.

Other types of knowledge emanate from *authority,* for example, when you believe your doctor's diagnosis of your ear infection and take the antibiotics he prescribed rather than the advice of a random blogger who suggests you put garlic oil in your ear canal. Therefore, **authority** is a form of knowledge that we believe to be true because its source is authoritative. Parents, teachers, and professional figures are some examples of these sources of knowledge. Knowledge also comes from experience, which is one of the first ways we learn as children. A child learns that it is dangerous to put hands on a hot surface because it can burn him or her. **Experiential knowledge** teaches us through pleasant or unpleasant experiences and continues throughout life.

Scientific knowledge, on the other hand, is based on studies conducted by researchers. In a nutshell, scientific knowledge is knowledge we can trust. It is through systematic research that we produce new scientific knowledge. It appears that scientific knowledge is not directly related to other types of knowledge, but we are all aware that tradition, authority, and experience may drive scientific research, at least theoretically. Conducting research does not simply mean following a specific method and obeying a set of rules. It also means embodying a different way of looking at the world, viewing it through two or more perspectives simultaneously. Sometimes it means gaining a fresh pair of eyes. So do we actually *know* reality? From the very start we must recognize that *reality and knowledge are two different things.*

By conducting research, we attempt to get closer to reality by attempting to build knowledge about it. But *reality* can be like an abstract concept that fades away every time we get closer. Like ants carrying bits of food, we march forward to find the truths we seek. Therefore, we can say that scientific research is the final product of conducting rigorous research. We generate this product by following a set of specific rules, embodying a set of specific skills, and embodying a specific framework when analyzing our results. This book will familiarize you with the discipline and fortitude of these hard-working ants, while simultaneously trying to instill in you the energy and the passion that it takes to become a great researcher. So let's have some fun!

Let's be honest, conducting research is not everyone's cup of tea. It is likely that you have plans for your future career that do not involve scientific research, so why bother

FIGURE 1.1 ■ Types of Knowledge

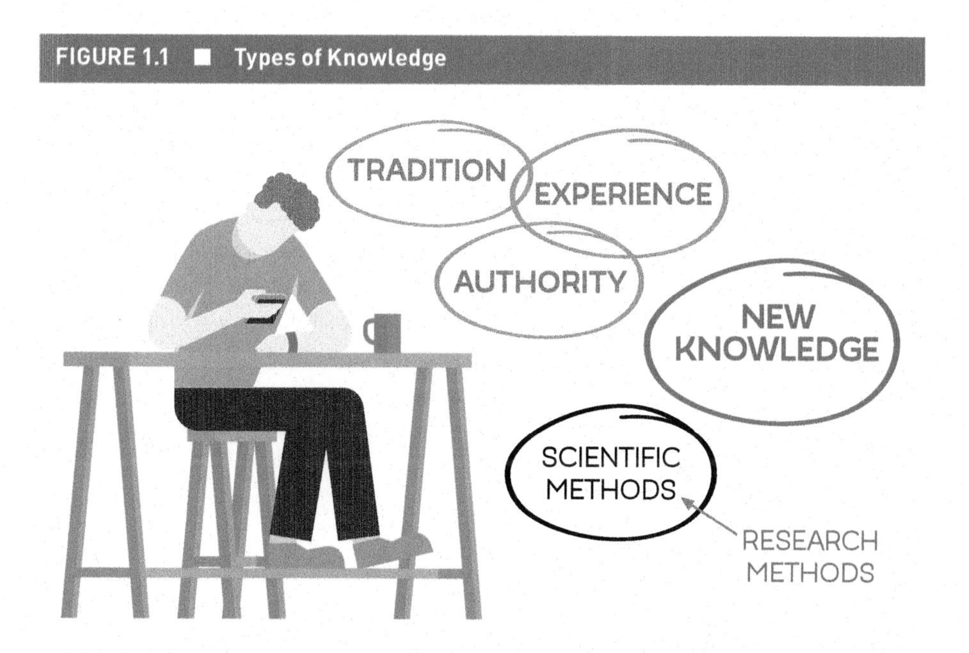

with this stuff? Here are three reasons that may change your mind. Note that none of them include "because it is required for your major."

1. Conducting research can be fun when you are in charge of your own work or study.
2. Knowing how to do research will open many doors for you in your career. It will open your mind to new ideas on what you might pursue in the future (e.g., becoming an entrepreneur, opening your own nongovernmental organization, or running your own health clinic), and give you an extra skill to brag about in your job interviews.
3. Understanding research will make you an educated consumer. You will be able to evaluate the information before you and determine what to accept and what to reject.

Imagine yourself in the supermarket trying to choose between the many types of apples in the store. Some apples are marked as "organic" and others as "conventional." There are also different types of apples that come in varying colors, are grown in different locations, and, of course, have different prices.

It is because of research conducted on the harms of pesticides used to grow conventional apples that you know the dangers of conventional products. It is also because of research that you are aware of what happens to the fruit when it is transported from thousands of miles away. Combining this knowledge allows you to decide what types of apples are the best for your health and budget. Though this example refers to something as simple as buying apples, we can use knowledge from research in all other aspects of our lives. Truth be told, understanding research will save you money in the short and long term.

In sum, we can conceptualize scientific knowledge as the kind of knowledge that follows detailed guidelines to reach conclusions. Scientific knowledge provides us with specific findings and information on how these findings became available. The "how" part is covered by the research methodology where we document all the steps we took to come to a new finding or new knowledge. But before we go into the details of methodology, we must take a peek into some theories of knowledge. Theories of knowledge attempt to explain in general terms how new knowledge is created and the philosophical approach for creating new knowledge.

THEORIES OF KNOWLEDGE

By conducting research, we develop and construct new knowledge. Many different theories have attempted to define how knowledge is created. The meaning of theory is further explained in Chapter 3, but for now, let us consider theory as a conceptual framework that we use to explain something around us. Theories of knowledge, for example, attempt to explain how new knowledge is developed. It is the reasoning behind creating and discovering new knowledge. Two of the most important and perhaps widely accepted of such theories are Karl Popper's falsifiability and Thomas Kuhn's structure of scientific revolution.

Karl Popper's Falsifiability

Sir Karl Popper was one of the greatest scientific philosophers of the 20th century (Stangroom & Garvey, 2015). His theory of falsifiability is a fascinating explanation of the growth of knowledge that we can apply to our daily lives and can influence the way we think and act. Popper devoted much of his thought and writing to the understanding of how knowledge grows and advances. His ideas are still applicable to today's research.

Popper observed that many grand theories claiming to explain everything about the world often err. What theory could be applied to absolutely everything that exists? Slowly but surely, he realized it was his systematic attempts to prove things wrong that advanced scientific knowledge. Let's illustrate this point with a simple example. If we know—the word *know* here is of key importance—that drinking coffee in the afternoon can keep us up later than our usual bedtime, we may refrain from drinking coffee when we plan to go to sleep as usual. On other occasions, we may want to drink one cup so we can stay up later to finish a paper that is due tomorrow. We take this knowledge for granted, and we apply it on a daily basis (i.e., drink a cup of coffee early in the morning, stay away from it in the afternoon).

One afternoon, we find that we are extremely tired. In fact, we are so tired that we could go to sleep at 5:00 pm and not wake up until the next morning. However, we don't want to go to sleep yet, so we get a cup of coffee even though it is late in the afternoon. To help ourselves feel energized, we eat some dark chocolate or a double-chocolate brownie from Starbucks, increasing the amount of caffeine in our bodies even more. Remember, we know that coffee will keep us up because this has been our previous experience (let's be professional here and call this experience by its scientific name: **empirical evidence**). Empirical evidence means acquiring data or information by systematically observing people or events. It comes from gathering data from practical experience.

However, this time, the caffeine in our body does not work as we had predicted from empirical evidence. Instead of energizing us and keeping us awake, it actually put us into a deep, dreamy sleep. We wake up 3 hours later, surprised that the coffee did not work. In Popper's terms, we have **falsified** an established theory. We have proven it wrong. By proving it wrong, we have added a new piece of knowledge to our already known theory. Now, instead of claiming that caffeine always energizes our bodies, we are claiming that sometimes—depending on how the body reacts to it—caffeine can have the opposite

Karl Popper

effect and put us into a deep sleep. We falsified an established theory and built a new theory on this knowledge. Falsifying a theory is our attempt to disprove an established theory, which is how we construct more advanced knowledge.

This is how we build new knowledge. Popper believed that in order to construct new knowledge, our goal should be to falsify the established theories. Advancing knowledge is an evolutionary process that he expressed through the following formula:

PS1 -> TT1 -> EE1 -> PS2

PS1 is a problem situation or issue that interests us or has a question attached to it. To explain this problem, there are a number of tentative theories, or TT1. If we try to falsify these tentative theories by error of elimination, EE1—or a process similar to natural selection—we find that most of our tentative theories are incorrect and there is a new explanation for the first problem we started on. Through this natural selection process, we build new knowledge and end up with a new problem situation, PS2.

FIGURE 1.2 ■ Popper's Tentative Theory Development Illustrated

To revisit our coffee example, we could say that our new problem situation, PS2, is that caffeine works sometimes to keep us awake, but not always. There are cases when caffeine will cause the opposite effect on our bodies and put us to sleep.

Thomas Kuhn

We end up with a new, stronger theory about caffeine and sleep. However, that does not mean this is an absolute principle. Rather, it is simply accepted until we succeed in falsifying it again. Popper brought to us a simple but important understanding of how knowledge is built, and this is how our everyday knowledge is constructed as well. We accept something as true until the moment we falsify it. Once we manage to prove it wrong, we build a better understanding on that particular theory or piece of knowledge.

Thomas Kuhn's Structure of Scientific Revolution

From early on in life, Thomas Kuhn was interested in the history of science and how knowledge is constructed. He defined knowledge as a summary of general truths and laws about the world that are scientifically proven. But how does science develop further? Most scientists occupy themselves with *normal science,* which, according to Kuhn, is basically what we know: general rules, general laws, paradigms we have accepted as truths, and so on. **Normal science** does not aim to explore new ideas, to build on scientific knowledge, to experiment, or to risk. It functions on what is already known and uses what we know as the ultimate truth. Normal science is SAFE. It is doing what we have been doing: relying on existing knowledge and not testing it.

This reminds me of my husband's cooking habits. He will follow a recipe to a tee. If one ingredient is missing, he becomes paralyzed. If the recipe calls for onions, and instead we have leeks—a cousin of onions—he will never use leeks. In Kuhn's terms this is normal science.

Normal science is made of accepted **paradigms**. A paradigm is an unchangeable pattern that we use over and over again. There are specific rules and regulations governing the paradigm that are widely accepted from a specific scientific community. Normal science is composed of many such paradigms, and we follow those in order to reinforce what we already know. Scientists who subscribe to normal science, according to Kuhn, will not discover anything new, just like my husband will never know whether compared to onions leeks work better, worse, or the same in the recipe. They are invested in reproducing the same normal science over and over again. Boring, if you ask me.

The following direct quote from Kuhn's (1962) *The Structure of Scientific Revolution* explains the paradigm in more detail:

> Paradigm is a term that relates closely to "normal science." By choosing it, I mean to suggest that some accepted examples of actual scientific practice—examples which include law, theory, application, and instrumentation together—provide models from which spring particular coherent traditions of scientific research. . . . The study of paradigms, including many that are far more specialized than those named illustratively above, is what mainly prepares the student for membership in the particular scientific community. (p. 8)

Every now and then, we encounter **anomalies**, or things that do not fit into the paradigms of normal science. When these anomalies occur, our understanding of normal science shatters. An anomaly is something that happens once or twice that does not fit into our commonly accepted patterns. When these anomalies start to occur left and right, they are no longer anomalies, but a **crisis**. A crisis is further defined as the accumulation of many anomalies against an accepted truth or normal science. We encounter a crisis when the normal science does not seem to fit with reality any longer.

Sometimes, this makes me think of the prevalence of mental disorders in our society. Often, mental disorders, as defined by the *Diagnostic and Statistical Manual of Mental Disorders,* refer to behaviors that are abnormal—meaning that they are different from what is widely considered normal. Now, if the number of such abnormalities increases all the time, we may need to reconsider the definition of what is truly normal. According to the Center for Behavioral Health Statistics and Quality (2016), an estimated 43.4 million U.S. adults aged 18 or older have some form of mental illness. This represents 17.9% of the population of U.S. adults. If almost 20% of the population exhibits behaviors that deviate from what is defined as normal, then maybe we are encountering a crisis and need to reevaluate the definition of normality. Maybe we should conclude that having a disorder is the new norm rather than an anomaly.

When anomalies accumulate, we start questioning what we have accepted as truth, or normal science. Kuhn believed that in order to change established paradigms, we must undergo some form of crisis. Crisis can lead to a **revolution** of science. The revolution replaces the old paradigm with a new paradigm. This is when we see a **paradigm shift**. So a paradigm shift happens when the widely accepted paradigm encounters many anomalies that lead to a crisis, then a revolution, and finally settle into a new paradigm. There is a specific note from Kuhn (1962) in explaining the revolution of science that concerns all young researchers: "Almost always the men who achieve these fundamental inventions of a new paradigm have been either very young or very new to the field whose paradigm they change" (p. 90). This quote should encourage and excite you as you launch your first attempt to construct new knowledge.

Now, if you closely observe Figure 1.3, you will notice that we move from normal science, to anomalies, to a crisis, and to a revolution that gives way to a paradigm shift. However, the paradigm shift is connected back to normal science. It should not be forgotten that the new paradigm we built in response to the crisis and the revolution will soon be accepted and will become the new normal science. In Kuhn's (1962) words

> Scientific revolutions are here taken to be those non-cumulative developmental episodes in which an older paradigm is replaced in whole or in part by an incompatible new one. . . . The man who is striving to solve a problem defined by existing knowledge and technique is not, however, just looking around. He knows what he wants to achieve, and he designs his instruments and directs his thoughts accordingly. Unanticipated novelty, the new discovery, can emerge only to the extent that his anticipations about nature and his instruments prove wrong. (p. 96)

To illustrate this point, I will go back to my husband's cooking. He loves to cook a dish with shallots and chicken breast. The recipe involves putting a lot of shallots

FIGURE 1.3 ■ An Illustration of the Scientific Revolution

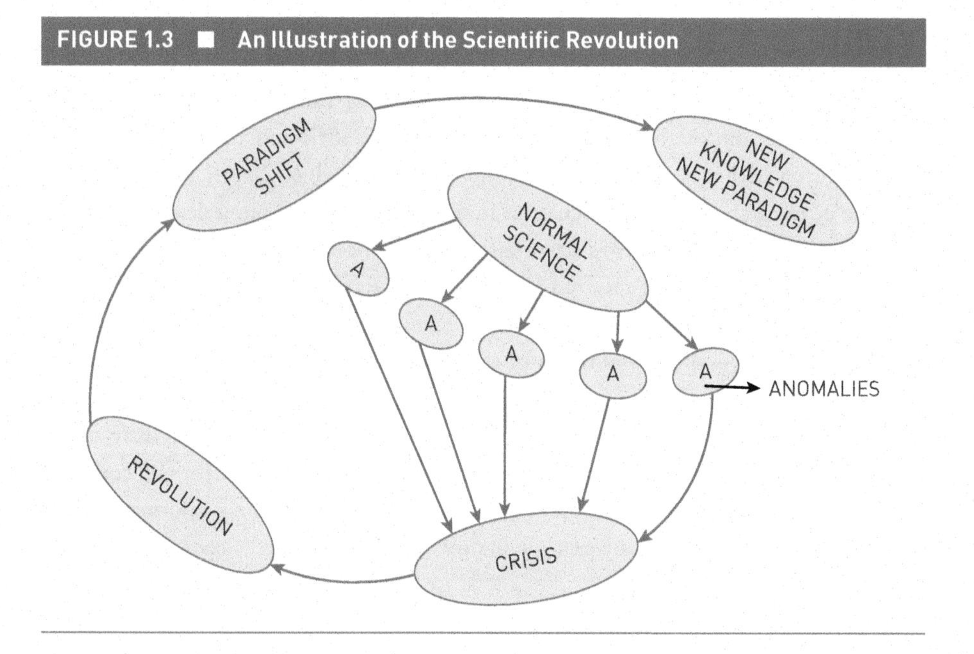

and garlic together in a pressure cooker. He adds some chicken breast, a bay leaf, black pepper, cinnamon, and one tablespoon of tomato sauce. It cooks for 30 minutes and, voila, becomes a wonderful, aromatic dish. We call it "çomlek," pronounced [chomlak] a popular dish in my home country of Albania.

One day, I was really craving çomlek, but there were no shallots in my pantry. I had everything else, but no shallots. I had some very large onions instead. I wondered how this famous dish would taste if I substituted the main ingredient with a related one. Shallots are, after all, onions of a different size. I peeled the big onions, partially sliced them, and inserted garlic in between the slices. I followed the rest of the recipe as usual, and waited for results. My çomlek was delicious! It tasted much better than the classic one because the garlic had melted inside the onion, giving it a very special texture and taste. After a crisis (not having shallots on hand) I caused a revolution and recreated the recipe with a different type of onion. The outcome was a delicious paradigm shift! Now, we never use shallots to cook çomlek. My new çomlek turned into normal science.

A QUICK LOOK AT QUALITATIVE, QUANTITATIVE, AND MIXED METHODS

There are three basic types of research methods: (1) qualitative research, (2) quantitative research, and (3) mixed methods. Qualitative research analyzes narratives in the form of words, texts, illustrations, videos, and other non-numerical formats. It is the type of research that requires deep interpretation and analytical thinking about what we are researching. It is almost an art form as much as it is scientific and allows researchers to express new creative ideas and innovative threads of knowledge. Quantitative research is the type of research that heavily relies on information retrieved numerically. It follows a strict methodology and requires attention to details prior to conducting the study and gathering data. Quantitative research allows us to try to falsify established theories and build a stronger knowledge.

Mixed methods is the combination of the two forms of research that allows for the flexibility to expand beyond one type of methodology and add information gathered by both types. Mixed methods provide us with numerical information as well as in-depth understanding of the data that were analyzed qualitatively. It is a favorite form of conducting research because of its undeniable strengths in providing additional information that cannot be covered by quantitative or qualitative studies alone. Let's see how they differ from each other.

Qualitative Research

Qualitative research aims at gaining insight and *depth* into whatever topic we want to know about. In qualitative research, we are not satisfied with simply drawing a picture of the facts; we want to know more insights, emotions, events, experiences, and details about the topic of research. We can be creative, connect issues, interpret the details we find, draw patterns from the raw information we collect, and so on. Surely, there are rules on how this is done, but the process is very exciting and highly creative. If we are studying people, we get to talk to them, hear their stories, find out their concerns, understand their issues, sympathize with them, and truly try to understand their actions.

A good qualitative study looks at an issue from various perspectives and attempts to detail a richer picture with a deeper understanding of people and events. We can use our artistic skills to describe what we have observed, bring out minutia that the common eye might miss, and direct attention to aspects of our research that no one thought about. Our work can be deep and engaging. In fact, there have been a few cases where researchers publish their work as a book and people enjoy it as fiction.

Conducting qualitative research means being immersed in the study and having a great deal of determination, attention to detail, and sense of commitment. Qualitative research requires a lot of contemplation on the topic in all stages, but especially during data collection and analysis. That is why qualitative research is more often based on what researchers call **inductive reasoning**. Inductive reasoning begins with specific observations and moves to a broader understanding of a topic or problem, which often leads to creating new theories of science. Inductive reasoning allows researchers to become immersed in their study without many preconceived notions or assumptions regarding how the results will look, but with the hope that many answers will be revealed as the work progresses. Therefore, being creative and interpreting the data collected are important parts of a qualitative researcher's work. Inductive reasoning allows researchers to shift the focus of their study as necessary during the process of data collection. In other words, researchers follow what they find interesting to investigate further.

Quantitative Research

Quantitative research starts with a lot of work up front, before the data are collected, and requires a good grasp of the topic of study and research conducted in that specific topic of interest. A quantitative researcher knows exactly what data are going to be analyzed, how the information will be collected, and even the types of procedures that will be used to analyze data. Their entire work is based on the systematic calculation of data. Researchers involved in this type of research are usually adept at designing and

RESEARCH IN ACTION 1.1

ILLUSTRATION OF A QUALITATIVE STUDY

In the following qualitative study, we can see how the author explains the qualitative methodology used and the rationale for using it. We can become familiar with the strict guidelines followed to ensure the highest quality possible.

Source: *Two paradigmatic waves of public discourse on nuclear waste in the United States, 1945–2009: Understanding a magnitudinal and longitudinal phenomenon in anthropological terms* by Judi Pajo (2016), published in *PLOS One*. http://journals.plos.org/plosone/article?id=10.1371%2Fjournal.pone.0157652.

FROM THE ABSTRACT

This project set out to illuminate the discursive existence of nuclear waste in American culture. Given the significant temporal dimension of the phenomenon as well as the challenging size of the United States setting, the project adapted key methodological elements of the sociocultural anthropology tradition and produced proxies for ethnographic fieldnotes and key informant interviews through sampling the digital archives of the New York Times over a 64-year period that starts with the first recorded occurrence of the notion of nuclear waste and ends with the conclusion of the presidency of George W. Bush.

From the abstract we become familiar with the scope of the study. This is qualitative research about the existence of nuclear waste in American culture. Let us take a look at the methodology where the researcher explains the qualitative methodology used.

FROM THE METHODOLOGY

The core of the sociocultural anthropology tradition, as it is commonly taught and understood, lies with the researcher personally going to an unfamiliar human community where the researcher spends a substantial amount of time, often in the range of several consecutive months. Known as participant observation or ethnographic fieldwork, this methodology consists of observing the broadest possible range of daily practices, in which, and for as much as possible, the researcher also personally participates. The purpose of such effort is to achieve an understanding of the world from the community's shared perspectives. The data collected through this traditional methodology consists mostly of a record of the researcher's own observations and impressions as well as statements from exchanges and interviews with the members of the previously-unknown community that, ideally, becomes better-known over the course of this process.

This is an excellent description of qualitative research and how it is carried out traditionally. Although some terms included here will be explained later in the text (such as *participant observation*), please note the importance of qualitative research to capture the community's perspective as well as the researcher's.

FROM THE FINDINGS

Nuclear waste continued to make headlines after 1969. Between 1969 and 2009, the New York Times reports that included in their headlines some combination of the keywords "nuclear," "radioactive," "atomic," and "waste" appeared amidst reports on political and environmental protests. The body of reporting identified here as "the second wave" corresponds to the presidencies of Richard M. Nixon (1969–1974), Gerald R. Ford (1974–1977), James Carter (1977–1981), Ronald Reagan (1981–1989), George H. W. Bush (1989–1993), William J. Clinton (1993–2001), and George W. Bush (2001–2009). The body of reporting since 2009 has not been included in the second wave, as the presidency of Barack H. Obama is currently ongoing. A total of 608 items of reporting are distributed by presidency as follows: 22 reports under the Nixon presidency, 18 under Ford, 157 under Carter, 160 under Reagan, 79 under the first Bush, 100 under Clinton, 72 under the second Bush. The core paradigm identified as characterizing this

(Continued)

(Continued)

wave is actually focused on nuclear waste: here nuclear waste appears to be a topic on its own right. This paradigm characterizes nuclear waste primarily in terms of the harm it causes, dissociated from the benefits of nuclear exploitation. The unspoken understanding appears to be that nuclear waste carries risks that cannot be eliminated, and that cleaning it up will involve costs that cannot be avoided. So instead of optimistic hope for a final and safe solution for the disposal of radioactive waste, this paradigm is preoccupied with assigning responsibility for radioactive waste.

Here the researcher is presenting one of the paradigms she concluded. Note how the keywords are identified at the beginning followed by illustrations of exactly where and how many of these keywords were found. After the information is given, the researcher provides us with the paradigm she identifies. We can see how the media primarily emphasizes the harm it causes rather than describing the benefits of nuclear waste.

using different means of data collection, a very difficult task, but necessary for collecting information properly.

Thinking ahead and taking measures for almost every detail of the study is the most difficult aspect of their work. In quantitative studies, a good sense of organization, categorization, and calculation is necessary before the study is launched. Quantitative researchers must be very specific on what they are testing for and put a great deal of work in preparing the most effective tool possible (e.g., questionnaire, survey) to measure the concepts and constructs they are targeting. Conducting this work in advance of data collection is crucial because once the study starts, there is little room for change. These pre-calculations and strong sense of organization give the quantitative researcher the ability to capture large amounts of data.

Quantitative research is commonly based on **deductive reasoning**. Deductive reasoning begins with a broad theory that can lead to a specific idea or concept that is ready to be tested. The researchers decide how and what measures to use in order to test the idea. This means that we have narrowed our focus of interest into measurable pieces and have specific expectations for the results of the study. Once the preliminary work is completed, we are able to collect our data. Data collection in quantitative studies is straightforward and there are no digressions or other routes taken by the researcher in the middle of the work. However, quantitative studies can be creative during the analysis of the data, especially if the researcher has collected enough information on various aspects of the population of interest.

RESEARCH IN ACTION 1.2

ILLUSTRATION OF A QUANTITATIVE STUDY

In this example, we can see how a researcher goes about sampling, data collection, and analysis of a quantitative study. This is a great example that can help familiarize us with the terminology used as well as the steps taken by the researcher to ensure high quality of data and the transformation of information into numbers.

Source: Food art does not reflect reality: A quantitative content analysis of meals in popular paintings by Brian Wansink, Anupama Mukund, and Andrew Weislogel (2016), published in *SAGE Open*. http://sgo.sagepub.com/content/6/3/2158244016654950 by Brian Wansink, Anupama Mukund, and Andrew Weislogel.

FROM THE ABSTRACT

Can the frequency with which a food is depicted in paintings give historical insight into family meals over the years and across countries? To initially explore this question, 750 food-related paintings were screened down to 140 paintings from Western Europe and the United States depicting small, family meals. Quantitative content analyses showed the most frequently eaten foods (such as chicken, eggs, and squash) were least frequently depicted in paintings. In contrast, the most aspirational foods such as shellfish were commonly painted in countries with the smallest coastlines (Germany), and more than half (51.4%) of the paintings from the seafaring Netherlands contained non-indigenous tropical lemons. Moreover, although bread and apples have been commonly available over time, bread has been painted 74% less frequently and apples painted 302% more frequently.

As we can see in this quantitative study, the researcher is analyzing the portrayal of meals in paintings from Western Europe and the United States. The abstract tells us that 750 paintings were screened to select 140 paintings of interest. Note that the researcher is also giving us some numerical understanding of the findings here as well.

FROM THE METHODOLOGY

The 750 food paintings originally identified were screened down to those focusing on family meals. After screening out paintings

(Continued)

(Continued)

of banquets, feasts, and still lives of food that were decorative and probably not full meals (such as bowls of fruit or game meat hanging on the wall), out of family meals that depicted food, a final total of 140 paintings from the years 1500 to 2000 were collected. Each painting was coded for all of the foods visible in the painting. The paintings were categorized by country as well as time period. The time periods were chosen as a rough representation of different periods in Western history. Paintings created from 1500 to 1650 were categorized as Era of European Exploration and Colonization, paintings from 1651 to 1850 were categorized as Era of Enlightenment, and paintings from 1851 to 2000 were categorized as Industrial and Post-Industrial Era.

At this part of the quantitative methodology, you can see how the sample of paintings was selected. It is important to note the amount of details about each step entailed in selecting the appropriate sample. The researcher is showing the guidelines followed for selecting the sample as well as how different periods were categorized with cut points.

FROM THE FINDINGS

Across years and countries, 19.29% of the paintings included a vegetable, 75.71% contained fruit, 38.57% contained a meat, and 41.43% contained bread. The vegetables with the highest incidence in the total number of paintings were, in descending order, artichoke, tomato, onion, squash, and radish, $\chi2\ (4, 2380) = 13.997$, $p = .007$. Similarly, the meats with the highest incidence in the total number of paintings were shellfish, fish, and ham, $\chi2\ (4, 2238) = 24.324$, $p < .001$. Shellfish, in particular, were depicted in over 22% of all the paintings, and they were most prevalent in Dutch paintings (56.76%) as well as in one fifth of the German paintings (20%). Fish and ham also had the highest incidences in Dutch paintings (13.51% each), although they were not as common as shellfish.

Even if some of the statistics are unclear to you at this stage, you should be able to see how the numerical findings are portrayed. We can clearly see how fruits occupied the highest percentage, followed by bread, meat, and finally vegetables.

Quantitative researchers often use structured questionnaires, surveys, or other types of questions with pre-defined multiple-choice answers, so it becomes possible to access information on a large number of subjects. To put this into perspective, we can think of the qualitative researcher as someone who knows a whole lot about one small, specific group of people and the quantitative researcher as someone who has access to fewer in-depth details, but the information available targets a much larger population. The qualitative researcher studies information in depth, whereas the quantitative researcher has a bird's-eye view.

Mixed Methods

Mixed methods refer to cutting-edge research studies that combine the best features of both qualitative and quantitative methodologies. Years ago this was not a viable possibility. The quantitative and the qualitative researchers were almost in opposing corners. But all of that has changed and researchers were quick to realize the great potential of combining these approaches. What are the benefits? One can collect a lot of information about a large quantity of subjects, in addition to much-needed details and depth on some of the participants. Mixed methods has become a desirable approach to research with excellent outcomes.

To better understand the benefits of a mixed-methods approach, let us look at a hypothetical example: Maia, a researcher, was interested in understanding whether eating cake was related to weight gain. She conducted a short questionnaire and surveyed many people at a local bakery. Maia asked them: (1) How often did you eat cake during a typical week this last year? (2) How much weight have you gained/lost this last year? (3) How often did you exercise during a typical week this last year? She wanted to see whether eating cake was related to weight gain in any way, and whether this was the same for each participant regardless of how much a person exercised. Maia also wanted to know more about the population she was interviewing. She randomly picked a few people to speak with at length, adding some qualitative work to her quantitative research and conducted some in-depth, unstructured interviews. When analyzing her data, she found that eating cake on a daily basis was related to participants' weight gain over the last year, but the relationship was not extremely strong. In some cases, the relationship was nonexistent regardless of whether people exercised regularly or not. It almost seemed like some people would lose weight and eat cake without exercising. This surely didn't make sense.

Maia investigated this topic further by analyzing her in-depth interviews. By talking to people, she realized that the participants who ate cheesecake daily rather than the ones who ate different types of cake gained the most weight. She also learned from these interviews that the bakery was strategically located next to La Leche League Clinic, which could indicate that a number of the women she surveyed had just had a baby and perhaps were still breastfeeding. Regardless of their sugar intake, they were losing weight from their previous pregnancy in conjunction with breastfeeding, instead of gaining weight. Maia was aware of the research findings from other studies that showed how breastfeeding was associated with weight loss for mothers. She then took another look at the quantitative data and saw that the majority of her participants were women. The combination of these details and insights on the quantitative information she gathered added depth to her study and explained why the relationship between cake and weight gain was not as strong as she had anticipated. Mixed-methods research led Maia to have more confidence in drawing conclusions.

ETHICAL RESEARCH

Now that we've gone over the basics of the types of research that can be conducted, we must discuss **ethics**. In daily use of the word, we may understand ethics as the group of morals and values that govern our behaviors and decisions. Deriving from this general understanding, we use a set of rules and regulations that are primarily concerned with protecting the rights of people who participate in research studies. These rules are called research ethics.

The ethical treatment of research participants is perhaps the most basic rule of conducting research. But what does this mean? Obviously, we should never do anything that may cause harm to participants. There are a number of ethical rules that must be followed to ensure that a researcher's participants are protected from harm.

Ethical Rules

One of the most important ethical rules is providing participants with information about the study, particularly about any risks that could be involved. Consider drug trials, for example. Pharmaceutical companies are eager to test new drugs on people. How else could they determine a drug's effectiveness? However, the drugs could have adverse side effects. Participants must be informed about these potential risks before participating in the trial. If English is not their first language, they need to be informed about these risks in their own language.

Furthermore, when a drug is tested for effectiveness, researchers use a second group of people with the same characteristics as the group of people who are testing the new drug. The second group of people is referred to as the control group. Without their knowledge, this group is given a placebo (a sugar pill or another non-pharmacological substance with no effects) instead of the testing drug. Therefore, the control group will not be aware of whether they took the medicine or the placebo and the researcher can measure the results from both groups. This way, researchers are able to compare the outcome of the drug they are testing between users and nonusers of the drug.

Participants are assigned to a control group or testing group without knowing which group they are in, so this needs to be clear to them from the start of the study. Sometimes patients are eager to try a new drug because of a problematic illness and must be informed that they may not receive the medication as they had hoped. In some extreme cases, ethics would dictate that a participant's condition is so critical that immediate medical attention is required and participation in the study would be ill-advised. In other words, researchers must be honest with participants and inform them about how the study will be conducted and explain any issues that may occur. This is the first and most basic rule in conducting research.

Another important ethical rule is confidentiality. Researchers go to great lengths to protect the identity of and data about their participants, from using fictional names to securing records to ensure that participants cannot be identified. We often conduct research with vulnerable populations, such as illegal immigrants, drug users, victims of abuse, and people with mental health issues who would be unwilling to participate were we careless about confidentiality. As a general rule, regardless of how sensitive the research topic is, we must always obey the rules of confidentiality.

Along the same lines is the rule of coercing. It is unethical to coerce people to take part in a study even if the coercion is subtle. For example, say a professor is investigating drug use among college students. Do you think it would be ethical to survey students who enroll in his or her classes? Students enrolled in this researcher's class may feel obligated to participate in the study because of the power that professor has over the classroom and they may participate unwillingly. In addition, students may not feel comfortable providing information about their personal drug use habits to someone who may judge their behaviors. Such cases can easily result in biased studies and are a violation of ethics. You can easily imagine how students may not provide accurate information to the researcher about their drug experiences in such a study versus a study where anonymity is provided. These responses will lead to inaccurate or biased study results because the researcher will collect inaccurate data.

Conflict of interest is also an ethical factor. It refers to the possibility that the study we are conducting may protect or be part of an agenda of some third parties. For example, if we were to conduct a study on the likability of the latest movie shown in theaters and we are sponsored for the study by the film studio that produced the movie, we may be prone to look favorably at the likability rates. The fact that the film studio sponsored our study is a conflict of interest that needs to be disclosed when we present our findings. Conflict of interest should not be confused with offering incentives (e.g., money or gifts) to people participating in a research study. Sometimes, in-depth interviews or focus groups require a few hours with the researcher, and some researchers offer incentives that show appreciation for the participants' time. It is not mandatory to offer incentives for participation and it depends on how much financial freedom is available to the researcher. Keep these rules in mind as you embark on your first research study.

RESEARCH WORKSHOP 1.1

COMPLETE A COURSE ON PROTECTING HUMAN RESEARCH PARTICIPANTS

Go to https://phrp.nihtraining.com/users/login.php to complete Protecting Human Research Participants, a free course that discusses how to protect the rights of research participants. (When you register, check the box that allows you to participate in continuing medical education credits.) This online training course takes about three hours to complete and consists of seven modules. The information provided is rich with details, definitions, and case studies. There are four quizzes that you will also need to complete that measure your understanding and knowledge of research ethics. You can reenter and continue the course at your convenience. After you have completed the modules and quizzes, you will have the opportunity to print your certificate of completion.

A Violation of Ethics

The Tuskegee syphilis study is an infamous case of an ethics violation in research. Between 1932 and 1972, the United States Public Health Service and the Tuskegee Institute conducted a study on the effects of syphilis on the human body. The researchers recruited 600 African American men to participate, but did not disclose the focus of their study to their participants, who were simply told they were being treated for "bad blood." Many believed they were receiving free health services from the government.

Two-thirds of the participating men had syphilis, and despite the fact that penicillin was validated as an effective treatment in 1942, not one of the men in the Tuskegee study received treatment. This continued for an additional 30 years, and was finally revealed through a leak to the newspapers in the early 1970s. Only 74 participants survived. Around 40 of their wives were also infected and 19 children were born with syphilis (U.S. Public Health Service Syphilis Study at Tuskegee, 2013). The tragedy of the Tuskegee study is measured not only in numbers directly affected, but in the lasting resentment it caused within the African American community.

Researchers' Biases

An important skill of being a researcher is the ability to study concepts objectively. **Objectivity** means perceiving something from different angles without personal preferences or judgments. Objective thinking is based on the facts of what has happened rather than our thoughts or emotions about it. Being objective is difficult and some may argue even impossible, but we can get close to it. One way of getting closer to being objective is our ability to recognize our personal biases and be aware of them. This awareness will allow us to guard ourselves from subjective thoughts and preferences and little by little we can get close to being objective. Biases are detrimental to our research and although we may never fully get rid of them, it is important to reduce them as much as possible.

Subjective thinking is based on personal emotions, experiences, and prejudices. All people are subjective in one form or another because we are molded by our unique personalities and backgrounds. One common type of bias is called **selective observation**.

FIGURE 1.4 ■ Illustration of Researcher's Bias

Selective observation happens when we are focused on a specific occurrence or a specific group of people instead of including an entire sample in our observation. It implies focusing on what interests us and, consciously or unconsciously, failing to notice other things that may contradict our theory.

Another common type of bias is called **overgeneralization**. Overgeneralization happens when we use a small number of cases to draw conclusions about the entire population. Similar to selective observation, we overgeneralize when we use something we have seen once or twice and believe that this is how "it always happens." But can we get rid of all our biases? Perhaps not. However, to brighten this gloomy answer, we can reduce our biases, though we cannot completely stop our subjectivity.

In our effort to reduce biases and reach objectivity, we conduct research according to widely accepted rules. We are our greatest enemy. Our misconceptions, assumptions, and preconceived notions of the world interfere with the research process. However, though on the surface a contradiction, it is because of our assumptions, creativity, and understanding of the world that we can deliver magnificent research. Your creativity and your imagination play an important role in conducting research. Without it, your work will be monotonous. You may be great at following the rules of the game, but you need your own brand of creativity to make the most of your research efforts.

Summary

This chapter discussed the ways we receive new knowledge and distinguishes among the different types of knowledge. Traditional knowledge comes from information we gather from our culture and social environment, particularly from the rules, regulations, and behaviors we learn as children. Authoritative knowledge includes what our parents, teachers, and professionals tell us about life, behaviors, and social circumstances. We learn from experiential knowledge where our behaviors are modified because of our experiences. Although these different types of knowledge tell us about how life works around us, they may not be scientific. Scientific knowledge is the type of knowledge we trust the most because it follows strict scientific rules of discovery.

You were introduced to two main theories of developing knowledge: Popper's falsifiability and Kuhn's scientific revolution. In Popper's terms, knowledge is advanced when we are able to disprove an established theory and falsify it. By building empirical evidence that contradicts an established theory, we can create a new tentative theory that is stronger than the previous one. Popper believed that we should always try to falsify theories in order to advance knowledge.

Kuhn's scientific revolution is conceptually similar. Kuhn saw the advancement of knowledge as a small revolution in itself. We have some accepted truths that he called normal science. Normal science functions on what is already known and does not occupy itself with exploring new ideas. Normal science also includes accepted paradigms. However, occasionally we encounter anomalies—things that do not fit into the accepted normal science. The more anomalies we encounter, the more likely we are to move toward a crisis. The crisis will bring a revolution and a paradigm shift. This paradigm shift will substitute the old normal science with the new advanced science.

The chapter introduced the three main types of research methodologies: qualitative, quantitative, and mixed methods. Qualitative research is based on inductive reasoning and begins with specific observations and moves to a broader understanding of a topic. It attempts to bring new insights and create new theories based on specific observations of a topic. Quantitative research is based on deductive reasoning or the type of reasoning that looks at a problem with specific expectations and assumptions about the results of the study. Deductive reasoning begins with a broad theory and applies it to a specific measurable problem. Mixed methods is a combination methodology that uses the best features of quantitative and qualitative research. Mixed methods allows for a better understanding of a problem and the ability to look at a specific topic both broadly and narrowly.

This chapter also introduced you to ethical considerations in research and how to conduct research while protecting participants in any study. Whereas some forms of protection are more obvious than others, such as not causing any intentional harm to participants, others are subtler. There are rules of anonymity and confidentiality at the core of every study. Other rules include not coercing subjects to participate in research or being careful about conflicts of interest.

Finally, this chapter brought forth the researcher's capability of being objective and the importance of perceiving a problem from different angles without allowing our personal preferences to take over. Objectivity is difficult to achieve, but we can train ourselves to reduce our biases by becoming aware of them. Subjectivity is based on personal emotions, experiences, and biases that are part of who we truly are. Subjective thinking sometimes causes selective observation or overgeneralization. Selective observation happens when we pay attention only to a few selective cases or subjects in our study rather than its entirety. Overgeneralization happens when we think that we can apply the same findings from a small group of participants to the society at large. Though subjective thinking has its flaws, it is also a channel of our creativity—an important part of being a researcher.

Key Terms

Anomaly: something that does not fit into the paradigms of normal science.

Authority: a form of knowledge we believe to be true because it comes from authoritative sources, such as parents, teachers, and professional figures.

Crisis: an accumulation of many anomalies against an accepted truth.

Deductive reasoning: reasoning that begins with a broad theory that leads to a specific idea or concept to be tested.

Empirical evidence: acquiring information by systematically observing people or events.

Ethics: a set of guidelines that are primarily concerned with protecting the rights of study participants and are mandatory for the researcher.

Experiential knowledge: a form of knowledge that we learn through pleasant or unpleasant experiences.

Falsify: prove that a theory is incorrect.

Inductive reasoning: reasoning that begins with specific observations and moves to a broader understanding of a topic or problem.

Mixed methods: research studies that combine the best features of qualitative and quantitative methodologies.

Normal science: the work of scientists using the general rules, laws, and paradigms that are accepted as truths. It does not explore new ideas or build on scientific knowledge.

Objectivity: perceiving something from different angles without personal preferences or judgments.

Overgeneralization: a type of bias that occurs when a researcher uses a small number of cases to draw conclusions about an entire population.

Paradigm: an unchangeable pattern that is used over and over again.

Paradigm shift: occurs when a widely accepted paradigm encounters many anomalies that lead to a crisis, then a revolution, and then a new paradigm.

Qualitative research: research that seeks to gain insight and depth on a topic.

Quantitative research: research based on the systematic calculation of data.

Revolution: when an old paradigm is replaced with a new paradigm.

Scientific knowledge: a form of knowledge based on studies conducted by researchers.

Selective observation: a type of bias that occurs when a researcher is focused on a specific occurrence or group of people instead of including an entire sample.

Subjective thinking: thinking based on personal emotions, experiences, and prejudices.

Traditional knowledge: a form of knowledge we inherit from our culture that includes information that we learned as children that is now part of who we are and how we behave.

Taking a Step Further

1. What is the difference between reality and knowledge?
2. What is research methodology and what do we need it for?
3. Can you think of examples that can illustrate Popper's falsifiability?
4. How does inductive reasoning differ from deductive reasoning?
5. What are some examples that can illustrate Kuhn's paradigm shift?
6. How does traditional knowledge differ from subjective thinking?

2

FORMULATING A RESEARCH QUESTION

CHAPTER OUTLINE

WHAT WILL YOU LEARN TO DO?

1. Choose a research topic
2. Explain how to operationalize research constructs
3. Describe the different types of variables
4. Formulate the various types of hypotheses
5. Create a visualization of a research question

CHOOSING A RESEARCH TOPIC

Here is a secret about researchers: They don't feel that conducting research is a job or a burden. They *love* the topic they are investigating. They love reading about the topic. They can easily spend hours analyzing data. Researchers are excited when they encounter something new, and they can talk passionately about it for hours. This reality check is important because typically no one mentions it. We often think of researchers as hermits, confined to their own worlds and invested in a boring pursuit. This image could not be further from the truth. Research is not a 9:00 am to 5:00 pm job. It crawls under your skin and becomes an obsession. It is almost like an obsessive video game where you cannot rest until you reach the next level.

This discussion of love and passion in research may have gotten you thinking about what you *love* to read about. Ask yourself the following questions: What topic keeps my attention? What blogs or newsfeeds am I likely to read? What do I repeatedly search for online? What topics peak my curiosity and inspire endless discussion? Is it a new high-tech gadget? Is it fashion? Is it Facebook and what your friends are posting? The answer can sometimes be found at your doorstep, but if you are still wondering what you are passionate about investigating, it may help if you take a day off and simply observe your behavior. Note what news headlines catch your eye and what keeps you alert and curious. By the end of the day, you will certainly have an idea about what topic interests you.

Deciding Between Fundamental or Applied Research

Now that you have an idea about what you want to research and explore on your own, it is time to categorize whether your research is fundamental or applied. **Fundamental research** looks at the world at large and tries to generate new ideas or explanations about how the world works and why. This type of research aims at collecting information about large groups of people and may not have an application in our everyday life, at least not in the immediate sense. We need fundamental research to gain insights into our world. Both theories of knowledge from Chapter 1 are types of fundamental research. They don't have an immediate application to our lives, but they show us how the knowledge is developed. We then use these broad explanations as a basis to develop practical guides that apply to various aspects of life.

Another example of fundamental research is theories of migration. People move from one country to another in search of a better life. Many researchers have tried to determine why people move from one place to another and what triggers a mass exodus. Sometimes the answer seems clear—economic or political reasons (Mahler, 1995). Other researchers claim that there are some additional reasons related to social status (Pajo, 2007) that put global migration into motion. These theories are fundamental because they are not applied directly to people who migrate. Researchers do not intervene in their subjects' daily lives or bring changes that concern them, but they aim to gain a deeper theoretical understanding of the human condition.

On the other hand, **applied research** seeks to solve a specific societal problem or uncover more information about a particular issue. This type of research explores why people behave in a specific way. By understanding why, it allows us to find a solution to the problem. This has direct implications in practice and it increases our understanding of how things work.

For example, a researcher was always curious about reasons behind the growing number of children diagnosed with attention deficit hyperactivity disorder (ADHD) in our society. Let's imagine for a moment that this researcher conducted an experimental study and compared children in a very active classroom (where they learn by doing and participating) with children in a traditional classroom (where they learn by passively absorbing information brought forth to them). The findings showed that traditional classrooms had an increasing number of children with ADHD-like symptoms, but there were almost no children with attention or behavioral problems in the active classroom.

If everything else was equal (age, marital status of parents, socioeconomic status of parents, race and ethnicity, and parents' education), this researcher could state that perhaps by placing children with ADHD-like behaviors in an active environment, we could lessen the problem in a much more effective way than by prescribing medications. Restructuring our entire education system may be the solution. The findings in this hypothetical case have a clear application in society and suggest a solution to a common childhood problem. Studies of this type are used to test forms of interventions or training to evaluate their effectiveness in practice.

Narrowing the Research Topic

Now that you've decided on a fundamental or applied research approach, let's examine the steps needed to more narrowly focus your topic of interest. Figure 2.1 shows how we start with broader interests or topics, and by asking questions, we can narrow our topic to something more specific. There are two features of your broader topic that can lead to your research question: (1) the constructs of interest, (a.k.a. your topic,) and (2) the population of interest. Narrowing down by the constructs of interest is easily done if you start asking yourself a couple of questions. If you refer to your topic in a broader term (e.g., the society, health, well-being, education, and others), try asking

So how do we narrow down our interest to something specific?

yourself what exactly that word means to different people. You will notice that general terms often have various meaning from different people. So how do you narrow down your term so that when you mention it to a person who is unfamiliar with your interest, they will understand exactly what you mean? If you say to a person, "I am interested in the way society makes decisions," they may be unclear what you mean, or worse yet, interpret it completely differently from your intentions. Are you interested in how people decide what type of property they buy, or how they decide what movie to watch, or what type of food to buy, or what type of medical therapy to chose when they have a problem? Trying to be as specific as you can about what exactly you are going to study and the constructs of interest will help you narrow it down to just one sentence that will clearly convey what your study is about.

The second important feature that helps narrow down your topic is your population of interest. Let us assume for a moment that you have cleared out the first part of constructs. In this hypothetical example, you are interested in the *quality of education in the elementary schools in the United States*. Most people are expected to have a clear idea of

FIGURE 2.1 ■ Moving From a Broad Topic to a Narrow Topic

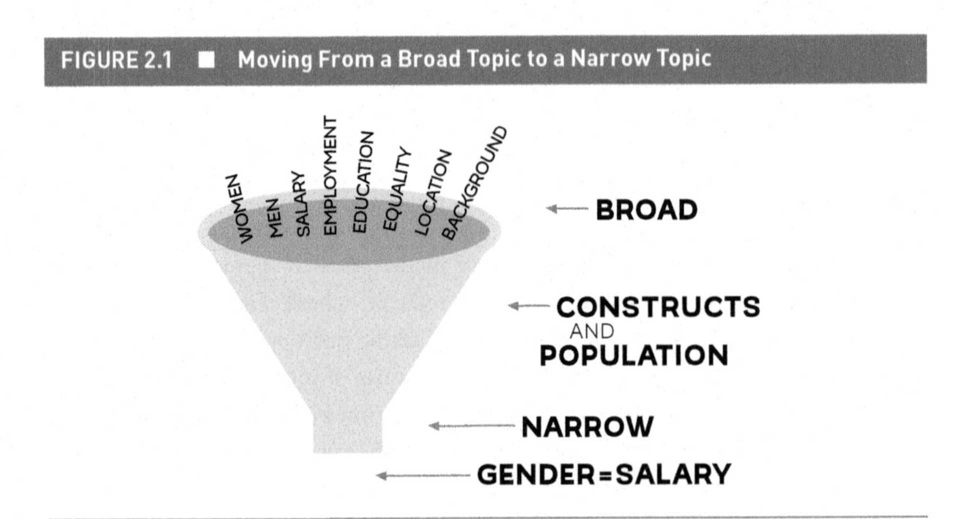

what we mean by this. However, as you may already notice, there are a few vague things in here that start with the word *quality*. Does quality of education mean how well students perform in elementary schools? Does it mean the quality of teaching? Does it refer to how parents feel about the quality of education? Here, the population of interest can clear out any inconsistencies. So we ask ourselves whether we are interested in students, parents, teachers, or other authorities involved in the elementary schools, such as psychologists or social workers. We may decide that we are interested in how parents feel about the quality of education in elementary schools, more specifically public schools. We can even go further and decide that by *parents* we actually mean fathers because their perspective seems to be missing from the current literature on the topic. Great. Now, we have narrowed down the constructs of what we want to study as well as the population of interest. Our topic now becomes: *fathers' perspectives about the quality of public education in the elementary schools in the United States.* Now, we can expect most people to have a very good understanding of the subject of our study.

RESEARCH WORKSHOP 2.1

AN EXAMPLE OF NARROWING DOWN A RESEARCH INTEREST

In this example, a student is trying to narrow down her topic by asking herself questions and answering them. Her interest is in health.

Question a: what type of health?
Answer a: children's health

Question b: what type of children's health? Lungs, heart, brain?
Answer b: no, maybe mental issues like ADHD, oppositional defiant disorder (ODD), or autism

Question c: So these are emotional and behavioral problems of children. What aspect of these issues is interesting to you?
Answer c: I am not sure how to answer . . .

Question d: Let's browse some aspects of the problem. Is it about how these emotional and behavioral problems start? Is it about whether children are being medicated for these problems? Is it about how the problems are identifying from parents and teachers? Is it about how parents experience these

(Continued)

(Continued)

problems emotionally? Is it about how psychiatrists diagnose these problems? Is it about whether some medications work and others don't? Is it about how disability centers in schools handle these issues? Is it about whether these children are successful in the long run? Is it about children who are identified with a problem, but are not taking medications? Is it about children who go to cognitive behavioral programs instead of taking medications?
Answer d: Hmmm . . . I think I want to look at how parents decide to go and check their child for emotional and behavioral problems . . . I mean, how do they know their child is even having a problem? What triggers the entire problem?

Question e: Great start since we know from previous research that there are no real blood or urine tests or x-rays that are conducted on children for this issue. Now, when you say children, do you mean preschoolers or school-age children? Can you narrow down by age?
Answer e: I had school-age children in mind . . . maybe elementary school children.

This student managed to narrow down the topic to this research question: "How do parents of elementary school children, ages 6–10 years old, decide that their child is having emotional and behavioral problems and need to ask for professional help?"

Try this type of exercise for your own topic by seeing if you can narrow it down to a very specific topic and by population. You will be delighted to see that you have a solid start for your research.

OPERATIONALIZATION OF CONSTRUCTS

Now that a specific research or problem has emerged, you may attempt to put the idea in the form of a research question if it fits the study you are conducting. Formulating a research question means a well-thought-out question that includes the gist of what your study is about as well as the population you are interested in studying. Depending on the specific topic of the study, you may also include the location where the study is conducted. The idea is to be as specific as possible so that most people would understand immediately what this study is about. Just as in the example above listing the fathers' perspectives on the quality of education, we need to convey our entire meaning of the study in one research question or research problem (if not a question).

The next step in finding out more about your topic of interest is finding good sources depicting studies of similar interest. The following short list will likely satisfy your search, but keep in mind that although these sources are a great beginning, they are not scientific sources for finding literature. Consider these more like friendly and easily accessible resources to get you started with new ideas on a specific topic.

a. Technology Entertainment and Design (TED) is an excellent source that includes videos of some of the finest and most elaborate research studies, unique ideas, and creative solutions. The website is free and friendly to navigate. Its focus is widespread in many fields, so it should be quite easy to find information on your topic of choice.

b. National Public Radio (NPR) is one of the best sources available. You can start by searching for keywords on your topic to see if it has articles, short stories, audio stories, or any other information. Its programs are well researched, made simple for larger audiences, and are likely to intrigue you into wanting to know more.

c. TED NPR is a collaboration of these two sources together in 1-hour weekly programs where a topic is investigated further by interviewing a few researchers who presented at TED conferences. There are many different topics available and this collaboration gives you various perspectives on the same topic at once.

d. The website of the Public Broadcasting Service (PBS) is another outstanding source for starters in any topic of research. Here you can find videos, long programs up to 2 hours, articles, and other rich information that again are very well researched, show both sides of any problem, and can get you excited about any topic.

e. The *New York Times* is perhaps the one source that will certainly have something in your topic. In most cases, your school library will have access to its online version where you can search articles on your topic. The articles here are well researched, written in an engaging style, and critically thought out. You will be happy you started here.

The next step in going deeper into your own research study will require digging into the scientific literature and getting your hands on articles about research that has been conducted on your topic of interest. This step requires a trip to the library or familiarity with the library databases. Here you will look specifically for research articles on the exact same topic you are hoping to conduct your study on. Creating an idea on what your topic will be about from the nonscientific sources recommended above is an excellent first step that will help you develop a more solid understanding of your own topic. When you have the idea quite clear in your mind, you can then look for scientific articles. The process will be much easier than delving into scientific studies without a clear research focus.

Operationalization is nothing more than our ability to transform constructs into measurement pieces. The best researcher is able to measure all the characteristics of a construct.

Once you have some ideas about your topic, it is time to talk about **operationalization**. Operationalization refers to turning our constructs into actual variables that we can measure. A **variable** measures a specific feature or aspect of your construct and can take different values. We need to return to our topic of interest and see if we can conceptualize our constructs into measurable variables. **Conceptualization** of constructs is the process of breaking down our constructs into smaller pieces and clarifying those pieces so we know exactly the meaning of each piece of each construct. Conceptualization of constructs is the first step toward operationalizing them. Operationalization refers to the concrete measurable variables of one specific construct whereas conceptualization refers to the breakdown of constructs into smaller pieces. Once we have these smaller pieces, we can then operationalize them or turn them into specific variables.

Say your topic of interest, or construct, is self-esteem. **Constructs** are general or abstract terms that are not straightforward to measure and are often understood differently by different people. If you want to operationalize self-esteem, you must figure out a way to measure it other than asking people to rate their own self-esteem. You would need to ask yourself: "How do I measure self-esteem? What type of questions should I ask my participants to get a sense of their self-esteem?"

So what is self-esteem? Dictionary.com states that self-esteem is a "favorable impression of oneself or self-respect." Great! This means that by asking people about what they think of themselves, you should have a pretty good idea about their self-esteem. But how do you do that? Let's say you are going to ask people to choose an answer from the following statements: Do they (1) strongly agree, (2) agree, (3) disagree, or (4) strongly disagree. Since we are on the topic here, we should be careful about giving people the option of odd-numbered answers because people tend to stay in the middle, which can be difficult to analyze later on, depending on the focus of the study. We may want to draw conclusions from the responses and cannot do so if everyone answers neutrally. By providing only four options, we force people to pick a side—either agree or disagree—with no gray area. So we can ask people to rate these statements using a one-to-four scale.

A. I think I am bright and have a number of valuable skills and qualities. (This statement should measure how important or the level of worth one feels about oneself.)

B. Most days, I like myself very much. (This statement should measure the likability one feels about oneself.)

C. Most days, I feel like a failure and want to hide from people. (This statement should measure the level of feeling like a failure one feels about oneself.)

The answers to these three statements should give you a good idea about someone's self-esteem. So we looked at the definition of self-esteem in the dictionary. That definition indicated that a favorable impression of oneself or self-respect is what self-esteem is all about. Although there may be many different ways of measuring self-esteem, we are attempting to go in a straightforward way from the definition. Favorable impression may be expressed on how much one likes oneself—likability. Favorable impression or self-respect may also mean how important a person feels—worthiness. Finally, to be certain that we are measuring positive as well as negative feelings about oneself, we are including a third question that measures these negative feelings—opposite to self-esteem, feeling like a failure. This way we are making sure that we are considering not only positive feelings, but also negative ones. We are also taking a measure against participants who may skip reading our questions and answer all the same way.

This is a simple example of operationalizing your construct of self-esteem into three variables. The answers that you would get for each question make up your variables. Continuing with this example, let's say that we are trying to figure out whether self-esteem changed regarding gender. We also need to operationalize gender. You have probably guessed that we will measure gender by writing:

D. Select your gender: (1) female (2) male (3) other

For the sake of illustration, let's say we gathered data from 40 students—half were girls and half were boys. Table 2.1 shows the results.

The purpose of Table 2.1 is to simply illustrate an example where we have operationalized a construct and collected the information from participants. Although you may feel that now we are ready to analyze the information, there are some additional aspects we need to consider and define first. We need to consider ethics of data collection as well as be clear on what type of variables we are using to measure our constructs.

TABLE 2.1 ■ Self-Esteem Data

ID	A-Worth	B-Likability	C-Failure	D-Gender
001	1	1	4	1
002	2	1	4	1
003	2	1	4	1
004	2	2	4	1
005	1	1	3	1
006	3	3	2	1
007	1	2	3	1
008	3	1	4	1
009	1	1	3	1
010	2	1	3	1
011	2	1	4	1
012	2	1	3	1
013	1	2	4	1
014	1	2	4	1
015	3	2	2	1
016	1	1	4	1
017	1	1	4	1
018	2	1	4	1
019	2	1	3	1
020	1	2	3	1
021	3	2	2	2
022	4	3	2	2
023	4	3	2	2
024	4	3	1	2
025	3	2	1	2
026	3	3	1	2
027	2	2	2	2
028	2	2	2	2
029	3	3	2	2
030	2	1	2	2
031	1	1	2	2

ID	A-Worth	B-Likability	C-Failure	D-Gender
032	3	3	2	2
033	4	3	1	2
034	4	4	1	2
035	4	4	2	2
036	3	4	2	2
037	2	2	1	2
038	3	3	2	2
039	4	3	2	2
040	1	1	1	2

ETHICAL CONSIDERATION 2.1

OPERATIONALIZING CONSTRUCTS FEATURE

The ethical rules of research must be considered when operationalizing constructs into variables. Questions need to be formulated in a way that will not cause harm to participants. We need to be sensitive and consider how the participant will feel about our questions and whether the questions will trigger any type of psychological distress. For example, if we are asking victim survivors to relive their pasts and tell us how they felt, we may cause them emotional and psychological distress.

Sometimes, we may not be able to avoid those negative feelings among participants because of the study's focus. If we were to research rape perceptions among participants, our questions may cause victims of rape to relive their tragedy. In such cases, we need to inform participants at the beginning of the study that the questions include details or words that may trigger psychological and emotional distress and remind them that taking part in the study is voluntary and they can stop at any time.

TYPES OF VARIABLES

A variable is the collection of the same information from most if not all participants in your study. A variable includes only one piece of information, but this information is collected from most if not everyone who is participating in the study. To make this definition clearer, let us take a look at Table 2.1 where we collected data on self-esteem. Each column represents one variable. You can see that the likability variable has collected information from all the participants about how much they seem to like themselves. The collection of this single information organized in a column represents a variable. Gender is another variable in this small dataset. The variable of gender includes information about the gender of all participants. Now that we have a broad definition of what a variable is, let us consider some categorizations of different types of variables.

Independent and Dependent Variables

Our research inquiry was to find out if gender influences a person's self-esteem. Using this research question, we can identify two important types of variables: independent and dependent variables. The **independent variable** is the explanatory or predicting variable that explains the variation in self-esteem—in other words, gender. It predicts the variation in the dependent variable and is often a constant. The **dependent variable** is the outcome, or the surprise variable—what we want to find out from a specific study. It is influenced by the independent variable. In our study, all three variables that measure self-esteem (worth, likability, and failure) are our dependent variables.

Note that the independent and dependent variables are unique for each study, so the same variable can be a dependent variable in one study and an independent variable in another study—it depends on the formulation of the research question. For example, if we were to ask a research question to determine whether the amount of make-up women wear relates to their self-esteem, our independent variable is the amount of make-up worn and the dependent variable is the measure for self-esteem. But if our research question was trying to determine whether self-esteem relates to the amount of make-up women wear, our independent variable is the measure of self-esteem and the dependent variable is the amount of make-up worn.

Control Variables

Going back to Table 2.1, let's look at how each variable has different values, because some people answered agree, some disagree, and so on. However, we have little additional information about these participants. It would be helpful to know each participant's age, family income level, parents' marital status, or anything else that may influence their self-esteem. For example, a participant with a higher income level may also feel a higher level of worth compared to people with a lower level of income. If we did not measure the variable of income and looked only at the data as they are, we may have concluded that gender is related to self-esteem when, in fact, controlling for the variable of income may have resulted in a different outcome. When we control for something in research, we are simply saying that we have taken into consideration the possibility that a specific variable had an influence on the outcome. To put into context the above example, if we say that even after controlling for income, gender still had an association with levels of self-esteem, we are implying that for people with the same level of income, gender was still associated with self-esteem. Any variables that we use to control our results are called

control variables. Control variables are not directly related to the focus of the study, but are crucial for understanding the relationship between the variables of our focus.

Control variables help to minimize biases and provide more accurate findings. For example, say that all the boys in the study came from single-parent families or families with low socioeconomic status and you forgot to control for either of these by including additional questions about parents' marital status and family income. Based on your data, you can see that the girls have higher self-esteem than the boys. This statement, however, would be far from accurate because you did not control for the variables of socioeconomic status and parents' marital status. The participants' self-esteem may not be related to being male or female, but instead to their family background and the way they were raised. Therefore, controlling for other variables can become quite important.

Control variables are as important as the independent and dependent variables. Think about it for a second. If we were to collect data only on gender and self-esteem, anyone could dispute our findings by raising valid questions: (a) How do you know if all the girls in your study were also very good students whereas all the boys were poor students? Maybe that is the reason their self-esteem is so different. (b) Sometimes children who are raised by single parents have low self-esteem. How do you know if most of the boys in your study were coming from single-parent families? (c) Did you control for their socioeconomic status? Maybe most girls came from a higher socioeconomic status, which could boost their self-esteem compared to boys? You get the idea, right? Therefore, in addition to collecting information on your independent and dependent variables, you will need to collect information on the control variables for other characteristics that could potentially complicate your findings and devalue them.

RESEARCH WORKSHOP 2.2

HOW TO IDENTIFY CONTROL VARIABLES

Control variables are just about any variable out there that can manipulate your variables of interest and as a result manipulate your findings in undesirable ways. Although there are some common control variables that may influence just about any study about society, there are some other control variables that may require a little bit more thinking. The first group of widely used control variables may include variables such as income, gender, education level, nativity, and race and ethnicity.

(Continued)

(Continued)

The second group of control variables is specific to your study and harder to identify. For one study religiosity may be of importance, but for another study marital status may be crucial to control for. Here are some sample questions to ask about your own study that may help you identify possible control variables:

1. What could possibly influence your independent or dependent variable?
2. What could possibly influence the relationship between the independent and the dependent variables? It may be helpful to think of cases that are contrary to what you believe—How are these cases different? Are there some characteristics that these cases have in common? That may be a control variable for you to consider.
3. Ask friends and family about your study without telling them everything about your study, but just what you are trying to do. Most people will share what they know about the topic and will also provide hints about control variables you may not have considered.
4. Look at the research articles in scientific journals on the same topic. Most studies will make it clear in their methodology section how many variables they controlled for and list these variables. This is undeniably the best source for you to start with in preparing a good list of control variables.

Confounding and Disturbance Variables

There are other variables that you may not be able to control for that could potentially ruin your findings. There are at least two types of uncontrollable variables: confounding or intervening variables and disturbance or extraneous variables. **Confounding** or **intervening variables** influence the independent variable in such a way that the results from the dependent variable become untrustworthy. We can try to protect our study from intervening variables.

Let's say you want to study the influence of Facebook participation on socialization. You collect data from 40 people on how often they log into their Facebook account and how often they post a status update or a picture. You also ask these people about how often they meet a friend or acquaintance for coffee or lunch. Your results seem to show that there is little, if any, effect on socialization from people who seem to be active on their Facebook accounts. Then it dawns on you that everyone who participated in your survey was single. They were probably going on coffee or lunch dates. The information on their marital status—the intervening variable—has influenced your independent variable and, ultimately, your results for the dependent variable.

The other problematic type of variable is the extraneous or disturbing variables. **Disturbance** or **extraneous variables** usually lurk in the background and they can disturb the findings of our dependent variable. Disturbance or extraneous variables are certain common characteristics (i.e., hypothetically collected in a variable) of our participants that are misleading the findings of our study without our awareness. Although we call them variables because they include common features of participants, most often we have not collected the information of these variables and are not even aware of their existence. Sometimes, disturbance variables become apparent to the researcher, but at other times they may never identify them. Disturbance variables are not directly related to the independent variable. Unfortunately, these variables may be out of our control when we conduct a study. We could simply be unaware of them entirely.

Recall the researcher from Chapter 1 who was trying to find out whether people who frequented a specific bakery were also likely to have gained weight during the last year. Her study saw some relationship between eating cheesecake and gaining weight, but the relationship was weak. It almost looked like it didn't matter much how cake a participant ate per day. She had not anticipated the fact that the bakery was next to a breastfeeding center, and the majority of women frequenting the bakery were mothers who had just delivered a baby. Breastfeeding is known to help women lose weight. The fact that these women just gave birth is the extraneous variable that could have potentially changed the results of the study.

Moderators and Mediators

There are two important types of variables you will likely encounter in your research: moderators and mediators. **Moderators** are variables that can strengthen or weaken an already established relationship between the independent and dependent variables. They are powerful enough to make a relationship seem weaker or stronger than the relationship is on its own. But the relationship between the two main variables is still there.

Let's assume that we want to investigate the relationship between alcohol consumption and liver damage. The amount of alcohol consumed daily would be our independent variable, and liver deterioration would be our dependent variable. Our dependent variable in this example would be determined by tests that measure the levels of specific enzymes and proteins in the liver.

After conducting this study, we find out that there is a direct relationship between the amount of alcohol consumed daily and liver damage. But we also see that this relationship is much stronger for women than men. The relationship is in fact weaker for people who have less body fat and exercise regularly. Although the relationship between alcohol consumption and liver damage is always clear, it seems to be stronger or weaker depending on these two other variables. Therefore, gender and weight are our moderating variables, because they moderate the strength of the main relationship in our study. So if you are male, in very good shape, and exercise regularly, you could binge drink and not have liver problems. Is that what we are saying? Well . . . not really. The relationship between alcohol and liver problems is still apparent, it is just weaker in this case.

Mediators are intervening variables that interfere with the relationship between the main variables. When a mediator is present, the relationship between the independent and the dependent variable may not even exist anymore. Mediators are strong enough to completely destroy a relationship between the main variables. Therefore, they are extremely important when we design our study.

Intervening variables are often called mediators because they are powerful enough to radically change the relationship between the independent and the dependent variable. For example, there is a well-known relationship between the amount of time a student studies and their grades; the more hours a student reads and prepares for classes, the better their grades. If we were to survey 100 students about the amount of work they put in and their grades, we might find out that the amount of work does not really match their grades. What could have happened?

Prior to conducting the study, we met with the school director and explained our student survey. The director misunderstood our study and thought that the amount of time students spent working was a reflection of the teaching at this school. Therefore this director was active in prepping students over an entire week so they would report a much higher amount of time as their study time. The instructions and prepping given from the director is our mediating variable that ruined the relationship we were measuring. Though this a hypothetical example, similar cases happen all the time, so we need to make sure that there are no extraneous variables interfering with our work.

TYPES OF HYPOTHESES

Now that you have identified a research question and operationalized it (operationalization is crucial regardless of whether you are conducting a quantitative or qualitative study) and you have a clear understanding of independent, dependent, control, confounding, and disturbance variables, it is time to take a deep breath and relax for a second. The time has come to think about whether you have a hypothesis that answers your research question or if you have a vague idea of what your answers could be.

Say that you would like to study eating disorders in young girls. If you believe that eating disorders are related to or influenced by the media's portrayal of female body image, you may guess that girls who watch more television are more likely to have some form of eating disorder. If this is the case, then you are stating a hypothesis. A **hypothesis** is a statement that predicts a specific phenomenon or behavior. In other words, a hypothesis makes a prediction about how people will respond to your research question. It can even go further in determining exactly how you expect the variables to behave after you have collected your data. Using a hypothesis, we offer an explanation to our research question. For example, if we were to question whether television is associated with women's self-esteem, we would be stating a hypothesis. But we can go further and state how we think the variables will behave by predicting that more hours of television watching is associated with lower self-esteem in women.

Alternative Hypothesis and Null Hypothesis

A hypothesis may seem like a guess as to what our results will be once we have measured all the variables, but it is more of a prediction that we need to test through statistical analysis. Rather than calling it just a hypothesis, we refer to it as an **alternative hypothesis**. An alternative hypothesis is therefore our possible prediction—based on the literature and theory—about what our testing results will be. In more specific terms, the alternative hypothesis states exactly how the variables will look once we have collected the data. This hypothesis predicts the type of relationship between variables and even what may happen to one variable if another variable increases or decreases in value. A study can have more than one alternative hypothesis, and for each one, we will have one null hypothesis. The **null hypothesis** claims that there is no relationship between the variables of interest in our study. So how does that work? Let's take another example and say that you are interested in finding out whether listening to music is related to sport performance for college-age student athletes. (*Note:* Hn is the notation for any alternative hypothesis and can take different numbers, like H_1, H_2, H_3, and so on. The null hypothesis is always expressed by H_0.) Your hypotheses are the following:

H_1: Athletes who listen to music for 4 hours or more per day perform better than athletes who listen to music for 4 hours or less per day.

H_2: Athletes who listen to jazz for 4 hours or more per day perform worse than athletes who listen to classical music for 4 hours or more per day.

H_0: Music listening and sport performance are not related to each other for college-age student athletes.

Now, if the alternative hypotheses are meant to predict a relationship between variables you are exploring, the null hypothesis always states that there is no relationship between variables under investigation. The null hypothesis is crucial to any research

RESEARCH IN ACTION 2.1

ILLUSTRATION OF OPERATIONALIZATION OF CONCEPTS

The following article provides an illustration on how concepts are operationalized, as well as how we create and formulate research questions and null hypotheses. The study is focused on how self-concepts and academic achievements are influenced by gender stereotypes of young students in secondary schools.

Source: Igbo, Onu, & Obiyo (2015). Impact of Gender Stereotype on Secondary School Students' Self-Concept and Academic Achievement. *SAGE Open*, 5(1), 1–10. http://journals.sagepub.com/doi/abs/10.1177/2158244015573934.

The major purpose of the study is to investigate the influence of gender stereotype on students' self-concept and academic achievement in senior secondary schools.

From the purpose, we can identify the major constructs: (1) gender stereotypes, (2) self-concept, and (3) academic achievements. These concepts are further operationalized into variables. Note that the study is looking for a specific group of students (seniors in secondary schools), and it will also consider the location of the school.

The following research questions guided the study:

- **Research Question 1:** What is the influence of gender stereotype on senior secondary school students' self-concept?

First, the study will look at the effects of gender stereotypes on students' self-concept.

- **Research Question 2:** What is the influence of gender stereotype on senior secondary school students' academic achievement?

Second, it will investigate the association between gender stereotypes and academic achievement.

- **Research Question 3:** What is the influence of school location on senior secondary school students' self-concept?

Third, it will consider the school's location on students' self-concept.

- **Research Question 4:** What is the influence of school location on senior secondary school students' academic achievement?

Fourth, it will also investigate the school's location on students' academic achievement.

The following null hypotheses were tested at the .05 level of significance:

- **Null Hypothesis 1 (Ho_1):** Gender stereotype has no significant influence on self-concept of senior secondary school students.
- **Null Hypothesis 2 (Ho_2):** Gender stereotype has no significant influence on academic achievement of senior secondary school students.
- **Null Hypothesis 3 (Ho_3):** Location has no significant influence on self-concept of senior secondary school students.
- **Null Hypothesis 4 (Ho_4):** Location has no significant influence on academic achievement of senior secondary school students.

The null hypotheses all state that there is no relationship between variables.

Note that for every research question, there is a null hypothesis to be tested. In this study, we see research questions, but these could have been formulated as alternative hypotheses as well. The researchers have kept them as research questions because they are nondirectional hypotheses. They are not making assumptions on how the variables will behave after the data are collected.

study, because *this is the hypothesis we are actually testing with our findings.* By collecting and analyzing data, we attempt to reject the null hypothesis. (We attempt to falsify, remember?) Therefore, since we are attempting to falsify the null hypothesis, we have only two options: (1) reject the null hypothesis in favor of our alternative hypothesis, or (2) fail to reject the null hypothesis.

You are probably asking why we have two negatives in one sentence. Can't we just accept the null hypothesis? Wouldn't that be easier for everyone? We can *never* accept the null hypothesis, because our goal is to refute it. We are trying to build new knowledge, and to do so, we need to test the null hypothesis with the aim of rejecting it. If we fail to do so, we are simply failing to reject the null hypothesis—we are *not* accepting it as valid. To accept the null hypothesis, we would need to test its veracity and that is a different research project.

Directional Hypothesis and Nondirectional Hypothesis

In other words, in a study, we are trying to determine if there is a relationship between two or more variables. To do so, we attempt to test and reject the null hypothesis, which basically states that no such relationship exists. When we prepare our alternative or competing hypothesis, we can choose a directional or nondirectional hypothesis. A **directional hypothesis** would predict a specific course for your variables. For example, if your hypothesis states that younger people are more empathic toward endangered animals, you are stating a directional hypothesis because you are predicting your results will show that the younger someone's age, the more empathy he or she will show toward endangered animals. There is more to this hypothesis than claims about the existence of a relationship between age and empathy toward endangered animals—you are giving this hypothesis a direction (i.e., the younger a person is, the more empathic he or she will be).

A **nondirectional hypothesis,** on the other hand, has no direction, but simply predicts a relationship between two or more variables. Often, nondirectional hypotheses are expressed in the form of a research question because researchers are not making any assumption on how the variables will behave, but are investigating the possible relationship between variables. For example, if you say that perceptions of rape are different between girls and boys, you are not specifying how this change is happening. You are simply saying that boys perceive rape differently from girls; there is no clear direction about this difference, which makes it a nondirectional hypothesis.

Open-Ended Question

Last, but not least, you may not have any prediction about how the variables will behave and whether they are related or not. In fact, maybe you don't even have variables, but rather a few constructs and a lot of curiosity about your topic. Then you may be heading toward an open-ended question: a question that may be exploratory in nature. If you are asking "How do people react to a chaotic situation?" that is an open-ended question that will only reveal its answers by collecting more information from people you are studying. You may think of scenarios or short vignettes you would like people to read and see how they think they would react in that particular circumstance.

VISUALIZING A RESEARCH QUESTION

Once everything has started to make a little sense, the best thing to do before all our thoughts are buried in some unknown part of the brain is to put everything on paper. And I mean visually. Get a piece of paper and visualize your research question in a simple drawing. Your constructs could be little or large circles and your variable could be rectangles. Make sure you think about controlling variables. If you are assuming or expecting a specific change to occur on your dependent variable from the independent variable, show this relationship with an arrow that goes from the independent variable to the dependent one. If you have a directional hypothesis in your mind, use + for increasing and – for decreasing to show how you expect the change to happen.

Let's go through an example. A researcher is thinking that when people exercise a lot, they are also likely to be eating healthy food. In other words, the more one exercises, the healthier food choices this person makes. This is a directional hypothesis because it has a direction. Now, how does one express this visually? There are two big constructs here: (1) exercise and 2) quality of food. First, we draw two circles, one for exercise and one for healthy food. In terms of variables, we are saying that the more we have from the first circle (exercise), the more we will have from the second circle (quality of food). Let's connect the circles with an arrow that goes from exercise to food quality and put a + in the corner of each circle, indicating the direction of our hypothesis. See Figure 2.2.

© iStockphoto.com/Rawpixel

Never underestimate the power of writing thoughts and ideas down on paper, even when you are certain that they will change again. Once written down, they become alive and explore further possibilities.

FIGURE 2.2 ■ The Relationship Between Constructs and the Breakdown of Constructs

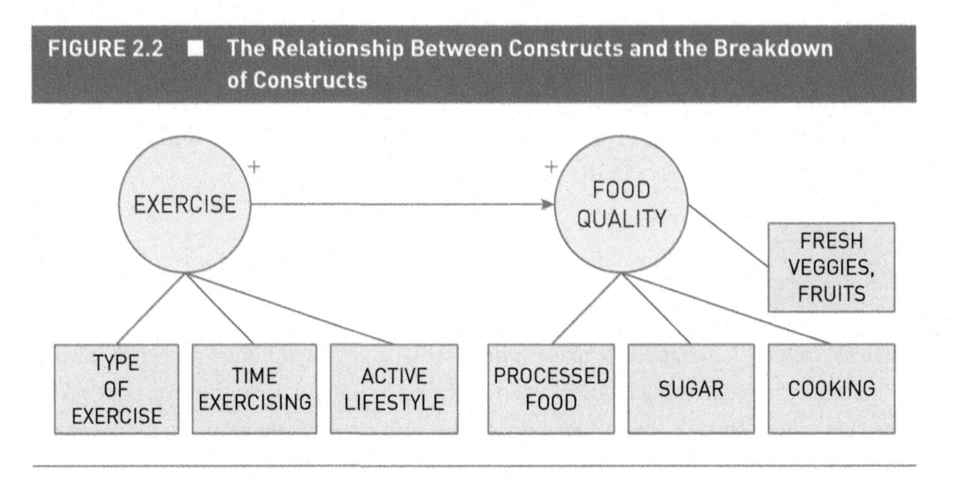

Let's take the first construct, exercise. How do we define exercise? Maybe you would classify exercise as a type of physical activity. Does the specific type of activity matter? Are we thinking of cardio or muscle building? It is better to have more variables, so let's keep this in mind: What type of physical activity do you engage in? Then we need to find out the amount of exercise, because we are hypothesizing that the more physical activity one does, the more likely one is to eat healthy foods. The amount of time one spends running, rowing, doing yoga, or walking is crucial. Therefore, we should ask: How many hours do you exercise per day? But wait a minute—shouldn't we ask if they exercise daily first? Yes! Our question is: How often do you exercise? Our multiple choices will have all the options (i.e., daily, biweekly, weekly, rarely, never). We could have the option that if people answer daily, they are asked how many hours per day.

Exercise is often determined by going to a gym and performing some form of activity. What about people who live in cities and walk or bike to work every day, walk up the stairs many times per day, or have jobs that involve carrying things, standing, or walking for hours? We may need to include another question that asks people whether they do physical activity as part of their daily routine.

Moving to the second construct, that of quality of food, we encounter a common problem in this type of research. What is considered good-quality food? This is a difficult question that is likely answered differently from different people. This is where investigating the literature comes in handy. If you haven't read the literature yet, you can still modify your model after you explore the research in this area. It usually helps to start from a tangible point and then modify it. You may be thinking that frozen dinners are not a healthy choice, along with sugar, chips, and soda. Returning to our model, how can we find out about our participants' eating habits? Perhaps you could make two lists of food that you identify as healthy and unhealthy and ask people to report how often they eat that product. Asking if they cook or not could be an additional good question.

Now, we have created some idea of what our model looks like and how we may operationalize our constructs. We will modify this model, but we needed a starting point. Finally, we need to add our controlling variables. What could influence the relationship we have stated between amount of exercise and eating healthy food? Generally speaking, what could make someone eat healthy or unhealthy regardless of exercise? Could gender make a difference? Could girls be more health-oriented than boys or vice versa? Let's control for gender. What else? People who work may have less time on their hands and opt for ready-made dinners or fast food. So we need to ask them if they are working or studying full time or part time. We also need to have an idea of their income. People with higher incomes may be able to afford better quality food compared to people who may not have access to organic or fresh food options. Your little graph of the constructs and variables is now ready and looks cool! Save this piece of paper (or file) because we will return to it once we have explored the literature and theories on the topic in which we are interested.

Summary

This chapter introduced the concepts of fundamental and applied research. We consider fundamental research those broad theories that attempt to explain how life works and can be applied to many different things rather than one specific problem. Applied research refers to studies that investigate one particular issue or problem. Once we have a topic of interest and have expressed this topic in the form of a question or a simple sentence, we are ready to operationalize our research. Operationalization means identifying the constructs of the study and expressing them in variables. Constructs are broad or general terms that cannot be measured straightforwardly, but need to be broken down to simpler measurements. Variables are simple, specific measures of one characteristic. A construct can have one variable or many variables.

There are a few types of variables. Some basic variables are independent and dependent variables. An independent variable attempts to predict the changes in the dependent variable. A dependent variable changes, or we expect it to change, as a result of the presence of the independent variable. Independent variables are known as predictors, whereas dependent variables are known as outcomes. Besides the main types of variables in our study, we also include other variables to make sure that our results are appropriate. Control variables, for example, are often not related to the main focus of the study, but they can influence the independent and dependent variables. To avoid biases and aim for higher accuracy in our work, we measure additional control variables.

Confounding or intervening variables may influence the independent variable and question the findings of our study. In our attempt to protect the study from confounding or intervening variables, we look out for possible variables that may influence our results. Disturbance or extraneous variables are more difficult to distinguish and take measures before the start of the study. These are variables that are not related to the independent variable, but they can potentially create a problem for the dependent variable, leading us to false results.

Moderators and mediators can also influence the independent and dependent variables. Moderators are variables that can strengthen or weaken a relationship between the independent and dependent variable. These are powerful variables that can make results much stronger or weaker, but the relationship between the main variables is still present. Mediators are even stronger. Mediators can cause the relationship between the independent and dependent variable to completely disappear.

This chapter also discussed the concept of hypothesis and its types. A hypothesis is a statement that predicts the relationship between variables. Some hypotheses can have a direction and make specific predictions about how the variables will change. These are called directional hypotheses. Other types of hypotheses do not make specific assumptions about how the change in variables will be reflected—they just state the possibility of an association between variables. These are called nondirectional hypotheses or simply research questions, if stated in the form of a question.

Another categorization of hypotheses divides them into alternative and null. Alternative hypotheses are statements that predict some form of relationship between variables, either directional or nondirectional. Null hypotheses are the statements that claim no relationship between variables. Null hypotheses are the hypotheses we test in the study. Even though it seems like we are trying to prove the alternative hypotheses, we are in fact simply testing the null hypotheses or the possibility that there is no relationship between variables.

Key Terms

Alternative hypothesis: the hypothesis that attempts to predict how the variables will relate to each other after the data are collected.

Applied research: research that seeks to solve a specific societal problem or uncover more information about a particular issue.

Conceptualization: the process of breaking a construct into smaller pieces and clarifying its specific meaning in our study.

Confounding (intervening) variable: variable that influences that independent variable in a way that causes results from the dependent variable to become untrustworthy.

Construct: an abstract term that is difficult to measure and can be understood differently by different people.

Control variable: variable that minimizes biases and provides more accurate findings by removing characteristics that could affect the relationship between the independent and dependent variables.

Dependent variable: the outcome or surprise variable that is influenced by the independent variable.

Directional hypothesis: a hypothesis that predicts a specific course for variables.

Disturbance (extraneous) variable: variable that can disturb the findings of the dependent variable, which cannot be controlled.

Fundamental research: by collecting information about large groups of people, research that tries to generate new ideas about how the world works and why.

Hypothesis: a statement that predicts a specific phenomenon or behavior.

Independent variable: the explanatory or predicting variable that explains the variation in the dependent variable.

Mediator: a variable that interferes with the relationship between the independent and dependent variables to the point where the relationship may be destroyed altogether.

Moderator: a variable that can strengthen or weaken an established relationship between the independent and dependent variables.

Nondirectional hypothesis: a hypothesis that has no direction, but predicts a relationship between two or more variables.

Null hypothesis: a hypothesis that claims there is no relationship between the variables of interest in a study.

Operationalization: the process of turning constructs into measurable variables.

Variable: measures a specific feature or aspect of a construct and can take different values.

Taking a Step Further

1. Find the dependent and the independent variables in these examples:
   a. A researcher is focused on how sleep (measured in hours) influences academic performance (measured in GPA).
   b. A researcher wants to find whether academic performance (measured in GPA) influences self-esteem (measured on a 1 to 10 scale).

c. A researcher is studying the influence of television watching (measured in hours) on speech onset of toddlers (measured in number of words spoken by age of 2).

d. A researcher is attempting to see whether self-esteem (measured on a scale of 1 to 10) relates to and/or influences substance abuse (measured in frequency of substance use and abuse) among adolescents.

e. A researcher is interested in understanding whether exposure to a large variety of food (measured in the number of different food textures and food types) influences the level of pickiness among children younger than 6 years old (measured on a scale of 1 to 10).

2. What is the difference between the null and alternative hypotheses?
3. What is the purpose of the null hypothesis?
4. How can we distinguish between disturbance and confounding variables? Illustrate with an example.
5. What are some ways to operationalize constructs, such as *sleep*, *time*, and *anxiety*?
6. What is the purpose of having a direction when we design an alternative hypothesis? What do directional and nondirectional hypotheses tell us?

RESEARCHING AND WRITING THE LITERATURE REVIEW

CHAPTER OUTLINE

WHAT WILL YOU LEARN TO DO?

1. Describe the purpose of a literature review
2. Learn about using libraries and online databases as resources
3. Create an annotated bibliography

4. Examine different ways of organizing a literature review
5. Discuss how to think critically and analyze studies
6. Identify the correct placement of study hypotheses
7. Create a shortened version of the literature review
8. Compare and contrast a systematic review of the literature and a literature review

DEFINING A LITERATURE REVIEW

The task of writing a **literature review** is a fulfilling endeavor, albeit a potentially daunting one. You may wonder where to start, how to continue, how to link the pieces of literature together, how to shape it, and, of course, how to cope with the fear of a blank page.

So what exactly is the literature review? The name is a misnomer—this crucial part of your research is anything but a *review.* The idea behind searching for the literature in your area of interest is that you become familiar with the entire body of research on that particular topic. It is expected that you have spent a lot of time reading these studies, pondering the details of each study, keeping abundant notes, accumulating arguments that support or criticize these studies, and systematically organizing them in terms of the time when they were conducted and the aspects they investigated.

You may now realize that becoming familiar with the entire body of the literature in a specific area may take a long time. However, for the purpose of understanding what a literature review is and how to properly construct one, you will be able to complete the task and get firsthand experience by using only a limited number of studies on your specific topic. To understand the ins and outs of a literature review, one needs to first conceptualize its purpose. Once that understanding is crystallized, the strategies of how to write it should come more easily.

Traditionally, there are two crucial points in the process where the novice researcher faces the scientific public at large: (1) defending the proposal and (2) defending the entire research study. The word *defending* here captures the traditional process used in academia when a *student* grows to become a *researcher.* Once the literature is investigated closely and the researcher is certain about the research design, he or she is ready to defend the proposed project. Scientists are invited to evaluate whether this researcher is ready to launch the new study. There are two tasks that the scientific community will be looking for during the formal defense: (1) whether the researcher knows the literature on that specific topic and (2) whether the proposed study is feasible. Writing the literature review addresses this first concern.

The new researcher needs to show familiarity with the literature and how his or her research will bring new knowledge or will solve contradicting areas in the field. But how is that done? The principle here is more important than the actual strategies, as various strategies exist that allow researchers to design the perfect literature review.

To make this simple, let's look at an everyday example. We know that by gaining knowledge, we can have a fulfilling life by succeeding in career pursuits and earning higher salaries. These days, it does not matter whether you graduated from a private or public institution or were self-taught as long as you have the necessary knowledge to be competitive in the job market. There are many different ways to gain the skills and knowledge you need, which is often a matter of personal preference and what works best for your goals.

The same goes for the construction of a literature review. It is important to show familiarity with the body of literature and to do so in an engaging style and elegant design, but you can be creative with how the literature is organized. What is most important, however, is to be able to think about the literature critically and shrewdly point out how your proposed study will answer questions from previous works. Looking at other studies with critical eyes is easier said than done. It is often helpful to look at a study's limitations section and understand some of the identified limitations that your study may be able to address. Other times it is helpful to find details about research that you may disagree with and construct a critique of these studies. At other times, it helps if you think about a study's conclusions and consider some ways that you would interpret these same conclusions.

Often we are misled into thinking that the body of literature includes only old studies conducted on the topic. That is far from the truth. The body of literature on any topic is a living thing, always changing and being improved on by new researchers who bring additional findings and information. If you conceptualize the literature as something alive that is eternally changing shape, you can think about how your own work will further modify and improve the current literature. It may help to visualize the literature on a specific topic and see how researchers have brought in new information that has transformed our previous knowledge on the topic over time. The way researchers design their literature review may be personal and most use their own unique style in constructing this piece, because writing the literature review is also an artistic endeavor. Prior to turning your literature review into an art piece, you need to consider the available studies and findings on the same topic.

EXPLORING THE LITERATURE

Sometimes we come up with a brilliant idea and are fairly confident that no one else has thought of it before. But the moment we sit in front of the computer and do a Google search on our idea, we are disappointed—someone else has already completed the same type of study we had envisioned. This should not be discouraging. We can learn a lot from understanding what others have researched on the same topic and even come up with something that no other researchers have studied before. However, we need to know what others have researched before we are able to formulate a truly original research project.

Exploring the literature will uncover related research and will help you to better formulate your own research questions. You will become more informed on the topic and your creativity may be triggered in considering new ideas that you may have missed the first time. It may also show that your initial inquiry has already been researched, but this will push you into new, unexplored realms. Do we really need another study that tells us that eating a lot of calories per day results in weight gain? Isn't this question overdone? A review of the literature may lead you to ask unexplored questions about metabolism of carbohydrates or proteins that would be exciting and worthwhile to pursue.

Exploring the literature can be a lot of fun. There is a sense of accomplishment in reading someone else's study and being able to say, "Hmm . . . I am not sure you really investigated this relationship fully, because you did not consider this or that detail." Ideally, you would keep notes of all the thoughts and ideas that come to mind as you are reading. Now, let's get into the nuts and bolts of how we explore the literature.

With available technology, we rarely take the time to visit our libraries.

Using Libraries and Online Databases

Your school library is your best friend and most valuable asset—besides your professors, that is. Not only do libraries have that wonderful book smell and friendly faces to welcome you, but there is something unique, refreshing yet old, about libraries. If you have not visited your school library except on orientation day, please do so. Do it for the spiritual enlightenment and the positive energy alone.

With available technology, we rarely take the time to visit our libraries anymore. Today, online databases provide almost everything at our fingertips. The starting point is your library's website. It is easily accessed from the comfort of your home or your favorite Wi-Fi coffee house. You may need a password or a username to access your library online. Once you are in your main library page, see if you can get to the link that takes you into "indexes and databases." Most libraries have these databases organized chronologically and/or by subject. The following is just a small sample of databases used for social and behavioral science, but there are many others available and they follow the same logic. Understanding how to use one database is the key to understanding almost all of them.

JSTOR I and II. This huge database is run by a nonprofit organization and has been around since 1995. JSTOR was built to help university libraries worldwide digitize their journals and books. It is a great source for any field or topic in arts and sciences. JSTOR includes over 2,000 journals and is widely used at over 8,000 institutions worldwide.

PubMed. PubMed is the largest database in the public health and medical areas and is operated by the National Library of Medicine. Its website advertises that it includes more than 24 million citations. It is a very handy database to use with specific keywords. It includes Medline and MeSH (medical subject heading) databases as well. It has built-in tours demonstrating how to use the database, and it is the first one to explore for any medical or health-related topic.

ERIC. This database is the best one in the field of education or education-related topics. It is sponsored and maintained by the Institute of Education Sciences of the Department of Education. It has over 1.4 million entries, and its navigation is user-friendly. Again, looking at its keyword system is helpful in order to determine your own key words or phrases.

PsycInfo. The American Psychological Association operates this large database. There are over 3 million entries in PsycInfo, and over 2,400 journals. This database is the first to use for any topic related to psychology or behavioral sciences.

There are many additional databases worth mentioning that you can browse and search on your own. This is a simple and limited list that provides you with some idea and initial sources to start with your search. Sometimes it is helpful to ask a librarian for help if you are having difficulties navigating through a database. Larger schools have specific librarians for specific fields whose job is to help you navigate the databases in your specific area of inquiry. The librarians in charge of a specific subject are up to date with new search keywords and new databases and are even able to assist you with your specific topic (e.g., social sciences will have a librarian, nursing will have a librarian, and so on).

Using Search Engines

Most databases have similar search engines, like Google Scholar, EBSCOHOST, or PROQUEST. They are easy to navigate and are designed with similar logic, and most students have no trouble navigating through them right away. However, there is one important thing to remember: always use "advanced search" rather than quick searches. Quick searches may list so many articles and records that they become overwhelming to explore. One can easily become discouraged and just walk away.

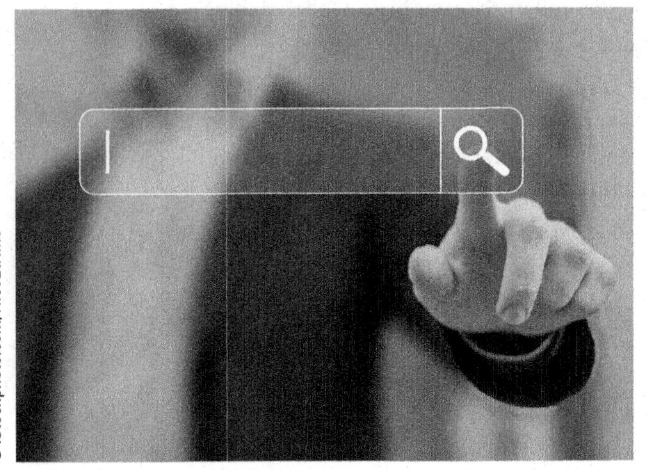

How to search is perhaps the first thing we learn to use in any new computer, laptop, phone, or any other gadget.

A helpful trick is to combine your keywords with **Boolean operators** (i.e., AND, OR, NOT). The idea is that you use AND when you want your keywords to be included together, OR when you want to include any of your keywords (OR will expand your searches rather than narrow them), and NOT when you want to exclude a keyword because you are not interested in the materials that include a particular keyword.

Another common trick is to use quotation marks (" ") when you want two or more keywords to work as a phrase together anywhere in the material searched. For example, you may be interested in "body image," but you don't want the search to list all results that include the words *body* and *image* separately because those materials may not be your focus. When you use quotation marks around "body image," the search results will list only those materials that have these words consecutively. Additionally, you can use an asterisk (*) to search for words that start similarly (this is called truncation). For example, if you are interested in all materials that include the words *child, children, childhood*, and *childish*, you can simply search "child*" and all the words that start with *child* will be searched.

The best place to start is to get used to some of the main keywords or subjects that these databases use. Most of them will have a link that provides their keywords or subject headings, which is one way of looking for a topic that interests you. Finally, make sure that you search for keywords in the abstract, not in the full text, because you may become overwhelmed with too many entries. If your search provides too few or no entries for your keywords, then you can expand to the full text and see if you have better luck. You may include other limitations, like the language used, or only listing full-text articles that you can immediately access through your library.

Using Interlibrary Loan

Let's reiterate the statement that libraries are your best friend. Librarians are able to find books and articles for you even if they are not available in your library's inventory. This is, by far, one of their most helpful features. You are able to access this inventory by using **interlibrary loan**. If you have never used interlibrary loan, you have missed out on a very convenient and often free service.

Let's say you find an article that may be perfect for your topic, but you cannot access the full version. Most schools should have a link on the article abstract page that says *interlibrary loan*. It is usually username and password protected (your school identification numbers should work here too, but some schools require that you create a different account for the interlibrary loan), and it takes you to a different page where you request the article. After you submit your request, a librarian will start searching affiliated libraries around the country to see which one has this specific article and will have the article sent to you via email. You can do the same for books, DVDs, or other materials of interest. In some cases, you might receive your article or book in a few hours. Other libraries may need a bit more time, depending on their connections with other institutions. This service allows you to access almost everything you need.

In addition to your school library, you may want to browse online for articles, books, or other resources. One user-friendly engine is Google Scholar. It may help to look there and see if it provides results that you could not find in your initial search, but full-text articles are usually not available from Google Scholar—you might be prompted to pay up to $100 or more per article, and you may not find as many sources as you would from your specific field database.

Depending on your location, you can also use public libraries. Some cities like New York City have gorgeous libraries—and that means more than just the building. All public libraries have online databases for articles and their own interlibrary loan system. Another great resource may be another university in your area. Some universities have larger libraries than others and even though you may not be enrolled in that university, most of them allow you to browse and read for free while on their campus.

Another resource for research studies on your topic can be found in the reference list of one great study. You may have to put in a lot of effort to find the one study that has to do exactly with the topic you were thinking about. As you read through it, you will notice how other studies are also mentioned in it. At the end of the article or book, there should be a detailed list of references. These references are often the easiest to access.

When you go to the reference list, you will find the name of the author(s); the journal (if the work is an article), including volume and issue number, or publisher (if the work is a book); the title of the work; and the page numbers where the work is found. You can go to the databases of your library and access the list of journals to see if the one you are looking for is available, or you can use that information if you need to submit a request via interlibrary loan as well.

Writing Annotated Bibliographies

One of the most common student questions about literature reviews has to do with the number of articles they need to read on a topic. It is a difficult question to answer and here is why. To conduct a real literature search on your topic you *need* to have read *everything* that was ever researched on the topic. You need to be familiar with all the prior research, including the names of the researchers to as far as their studies' strengths and weaknesses. It is essential to know what exists before you introduce new research that will add to the body of literature and be subject to critical opinion.

Research is an attempt to discover new information about a specific topic—it is a form of creative invention. If you were to invent a type of sugar from some root or plant or chemical mixture, for example, you would need to know all the types of sugar that exists before you consider investing your time and effort. Now, with that said, you may not have time to conduct a thorough literature search during the time available during a research methods course. Despite the time constraints, you must still try to read and absorb as many articles as possible on the specific topic, but it may be difficult to read every single piece of literature on your topic.

Once you have your articles, it is time to start reading. Most peer-reviewed articles may be challenging at first, but reading them slowly, sentence by sentence, should do the trick. Quality is more important than quantity so if you truly have a great research study on your topic, take the time to read it thoroughly, even if it takes an entire day. It is always helpful to create an annotated bibliography before you move to writing your literature review.

An **annotated bibliography** is a brief summary of the article or book you read, including the focus of the study, methodology used, conclusions, and anything else you found important while reading that particular work. It includes explanations or comments on the citations as well. As you are reading, keep notes with a piece of paper and a pen or your computer. First, write the article title, author's name, journal's name, volume number, issue number, digital object identifier (DOI) number, and the page numbers of the article. You do the same for books, videos, or other materials that you are exploring for your research. As you continue reading the materials, keep notes on information that is important to your own work and some general aspects of the study that you may need for referral. For example, you should note the following:

a. The focus/aim of the study: What was this study trying to accomplish?
b. The theory or framework that is mentioned in the study and that guided the research (this is often clearly stated, and if it is not, you may need to state that it was not mentioned)
c. How the researcher(s) addressed their research question/hypotheses: What type of design did they use (quantitative, qualitative, mixed methods)?
d. The number of people who participated in the study described in the article
e. The population breakdown in terms of gender, race and ethnicity, age, or other characteristics that may interest you
f. The results of the study: What did the researcher find out?
g. The main discussion points the author(s) write about and any future recommendations they make
h. Your own notes/impressions or anything that seems important to you or your own research, including quotes (cite page numbers when quoting the text)

The writing of annotated bibliographies does not have hard and fast rules, but is rather as if you are writing an email to a friend explaining the article you read. Do not worry much about the quality of your annotations just yet, but instead concentrate on writing down all the facts and preparing good summaries of your literature.

UNDERSTANDING AND ORGANIZING THE LITERATURE

There are two more steps to accomplish before we move to the ins and outs of writing the literature review. We need these steps to lay the groundwork for a successful and engaging synthesis of the literature. The first is creating a guiding table of the articles we have read. The second is going back to the graph we conceptualized in the previous chapter.

Creating a Guiding Table

In order to ensure that all the information from the articles is clear as you write the literature review, spending some time creating the most detailed table possible will

RESEARCH IN ACTION 3.1

ILLUSTRATION OF ANNOTATED BIBLIOGRAPHIES

Study 01: *Practice Guidelines and Parental ADHD*. . . . Regina Bussing and Faye Gary; 2001

Aim: Examining parental evaluations of treatment approaches to attention deficit hyperactivity disorder (ADHD).

Theory: None mentioned.

Methods: Researchers conducted four focus groups (two hours each).

Participants: 25 parents of different ethnicities (12 Caucasians and 13 African Americans); 50% are single parents. Parents talked about different measures in parental techniques, food choices, discipline, activities, religiosity.

Results: African American parents were more in favour of corporal punishment, religious practices, and strict discipline, whereas Caucasian parents were in favour of different forms of activities: rewarding and behavioral programs. Parents were concerned about medications. They felt that medications were freely given to them from pediatricians. Even though 38% of the quotes on medication from parents seem to say that medications are helpful, they are very skeptical about its use. Almost all parents said that they took their kids off medication during summer and holidays and even at home.

Personal Notes: In this article, it came up that parents put the blame on themselves. The calls from teachers were also brought up as a reason for medicating children. A distrust of professionals came up, however differently

For this writer, the existence or the lack of theory in the article is important, so you can see how there is a section discussing whether there is a theory in the article or not.

Note how each section of the article is separated in its own subsection for easiness. Also, the language in the article is informal, almost in a personal style to remind the writer of the main characteristics of the article.

from what I have read before. The distrust in this article was because professionals were medicating too quickly for parents' taste. Parents also reported having tried many things before going to see a specialist.

Study 02: *Motherhood, Resistance and ADHD.* Claudia Malacrida, 2001

Aim: Exploring mothers' perspectives on power dealing with medical, psychiatric, and educational professionals.

Theory: Foucault's theory of power.

Note that this article has a theory and the writer is making note of the specific theory used. This will make everything easier at the end when many articles are looked at together.

Methods: Qualitative methodology: in-depth interviews.

Participants: 34 mothers, 17 in Britain and 17 in Canada, none of them employed.

Results: The author reaches the conclusion that mothers follow a pattern of power that starts with resistance and follows with political actions and "playing the game," as they call it. Mothers in this sample express disappointment with teachers and psychiatric professionals. They feel they are being watched, talked about, or blamed as bad mothers or are considered pushy or overachieving mothers. They were blamed and ostracized. There was one incident where the mother proved this point, but most of the other cases are drawn from their own perceptions. There were a few important points in their interviews.

(Continued)

(Continued)

Personal notes: Wonderful article. I really like Malacrida's work. Again, receiving calls from the school came up. Mothers perceived that they were being the only ones blamed rather than the school or the teacher. How true that is, I am not to judge, but mothers in this sample do not seem to accept a combination of both. They see the problem as it affects them as if they are to blame. They also tried to be active in school and the way they described it was more as "a way of theirs to see what injustices were happening there to their children" (p. 7). Now, even if that is true and every school has injustices caused by teachers or other children, the fact that mothers try to participate for that reason makes them judgmental before they actually participate. All the mothers in this sample had the same conception of schools and teachers. The truth is that being a teacher is a very underpaid profession and one can imagine that teachers might not always want to have this occupation, so I can see how there is some truth in it. However, the fact that mothers are taking actions on their judgment, I wonder how their perspectives influence their own children. What happens to the child's ability to be responsible and defend him or herself instead?

There were also two instances to point out in this article. One was a mother who notes that she quit her job because taking care of a child with ADHD was too much for her. Additionally, Malacrida is the first who asks women how much time taking care of the child does it take in regard to other activities, and it seems that all women in the sample admit that these activities take up too much time on their ability to work on their desired careers, marriages, and lives of other children. This is so important for my study on parental time, because no one else truly mentions time like this.

Note how lengthy this personal note at the end of this article is in comparison to the first one. Clearly this is the article that has a lot of valuable information for the writer, and everything that is important or occurred to the writer while reading is noted in this section. See how questions are raised, relations to other information outside of the article, or even ideas on how this article adds to the entire literature on the topic explored.

help tremendously. Having a fancy design is a plus, but even a clean, straightforward table will do the trick. Just think of all the peer-reviewed, original research articles you read. Initially, all the data need to be extracted from these articles and synthesized in the literature review section. Creating this table helps you to understand these studies better.

There are various ways to organize this table, but one possibility is to start by having an identification (ID) number for each study you read in the first column. If it helps, you can write this ID number on the first page of the article (if you printed them). Assigning an ID number will make your life easier. The second column can include names of the authors. Depending on the topic you are researching, you may notice that most of the studies have been conducted and written by male researchers, or in other cases by female researchers—this type of detail may be important to note because it could be something to add in your paper.

The third or fourth column could be filled with the year the study was published and the year the study was conducted. Sometimes an article is listed as published in 2016, but the study was conducted in 2007. That information may be important to you once you look at the overall time of all the studies. In the fifth column, you may want to write about where the study took place. Studies are conducted all over the world, so depending on the topic, this may be important. For example, I was interested in understanding whether children were more commonly diagnosed and medicated for ADHD in the United States and Canada versus other countries, so the details about the location were important to me. The sixth column could be dedicated to the theory used in the article. If the article does not mention one, you could write "none." Most original research will likely have a sentence or two on the theoretical approach used.

The seventh column may hold information about the participants in each study (i.e., how many people participated, race and ethnicity, gender, education, socioeconomic, marital status, and anything else the article mentions about their participants). Your eighth column may include information on how the study was conducted (i.e., questionnaires, surveys, pre- or posttests, in-depth interviews, focus groups, or any other methods used).

You could also add a column on any other specific information that is important to your study. For example, if the topic of your study is adherence to medications for people with low blood pressure, you might dedicate one or two columns specifically to the information that relates to medication use. In addition, you would include a column on the results and findings of the study, one on the main discussion point, and a final column on your thoughts or notes or important quotes from the article. See an example in Table 3.1. You may add more details to your own table as you see fit.

TABLE 3.1 ■ Organizing the Literature

ID	Year Pub	Name	Author's Gender	Country Where the Study Was Conducted	Theory	Respondents	Types of Methods	Aim	Findings
01	1996	Robert Reid, Melody Hertzog, Marlene Snyder	Male, Female, Male	United States: Midwest	Grounded theory methods	20 parents (18 mothers and 2 fathers)	Semi-structured interviews (two rounds) 20–40 minutes and questionnaires	Parents' experiences in obtaining services for their ADHD children in school systems	Parents perceived schools as having failed to adequately respond to their children's educational needs
02	1997	Sarah F. Wright	Female	United Kingdom	N/A	17 parents (usually the mother): 56% both parents lived in the same house, 38% divorced, 1 case father had died	Semi-structured interviews	Parents' experiences of ADHD and Ritalin	Experiences vary a lot depending on what parents believe ADHD to be. Their idea of what ADHD is depends on their expectation and experiences with Ritalin
03	1997	Charlotte Johnston, Wendy Freeman	Female, Female	Canada	N/A	52 parents of ADHD children and 42 parents of children without behavioral problems. Mostly married	Questionnaires; there is a control group and this is a quantitative study	Comparisons of how parents of ADHD children behave and react compared to parents of non-ADHD children	Parents saw behaviors of ADHD children as uncontrollable by the child and internally caused, and

ID	Year Pub	Name	Author's Gender	Country Where the Study Was Conducted	Theory	Respondents	Types of Methods	Aim	Findings
						(for the ADHD group) and the majority married for the control group.		regarding the behavioral and emotional problems of their children	they saw themselves as not responsible for these behaviors
04	1998	Ruth Segal	Female	Canada	Grounded theory	17 families (12 dual-parent and 5 single parent) (3 fathers came along with 17 mothers), European descent, 9 mothers (including 5 single mothers) were employed	1–1.5 hours (two interviews with each family); interviews were 1 to 2 weeks apart. Interviews were audio taped and recorded; transcribed by a third party.	Exploring daily experiences of families of ADHD children	Homework and morning routine seems to be the hardest part of the day, and parents manage to cope using a great deal of organization
05	1998	Emily Arcia, Maria Fernandez, Marisela Jáquez	Female, Female, Female	United States: Dade County Florida	N/A	7 mothers, Cuban, 2 divorced and 5 married, average of 13.3 years of education, 5 worked and 2 didn't	Two sessions of interviews 13 months apart. Interviews were about 1 hour and audio taped. Plus structured questionnaires.	Exploring how Cuban mothers develop their ADHD schemas and understanding what motivates these schemas	Cuban mothers develop schemas with a combination of their cultural perception, academic achievement of the child, and authorities' opinion (teachers and doctors)

(Continued)

TABLE 3.1 ■ (Continued)

ID	Year Pub	Name	Author's Gender	Country Where the Study Was Conducted	Theory	Respondents	Types of Methods	Aim	Findings
06	1998	Judy Kendall	Female	United States	Grounded theory: Perspective of symbolic interactionism	59 respondents from 15 families: 15 mothers, 10 fathers, 20 ADHD children and 14 non-ADHD siblings, 14 Caucasian families and 1 African American; 5 single-parent families and 10 dual-parent families (3 cases a non-biological stepfather); 3 families had lower socioeconomic status whereas 12 others were middle to upper middle	109 open-ended interviews, 30–90 minutes in length and observations (individual and family interviews). Two rounds of interviews. Recorded and audio taped, observations.	Exploring how parents cope with behavioral disorders	Parents outlasted disruption of ADHD through four subprocesses of reinvesting, making sense, recasting biography, and relinquishing the good ending

ID	Year Pub	Name	Author's Gender	Country Where the Study Was Conducted	Theory	Respondents	Types of Methods	Aim	Findings
07	2000	Henrikje Klasen, Robert Goodman	Male, Male	United Kingdom: London	N/A	39 respondents: 10 general practitioners and 29 parents (19 mothers and 10 couples)	Semi-structured interviews, about 1–2 hours (audio taped)	Exploring the views that parents of ADHD children and general practitioners hold toward each other	Parents' and general practitioners' views of each other differed because of their ideas as to what ADHD was. Parents felt that doctors already had an idea as to what ADHD is and followed it without paying attention to individual cases, and doctors felt that parents didn't comply with whatever they thought of the disorder

(Continued)

TABLE 3.1 ■ (Continued)

ID	Year Pub	Name	Author's Gender	Country Where the Study Was Conducted	Theory	Respondents	Types of Methods	Aim	Findings
08	2001	Regina Bussing, Faye Gary	Female, Female	United States: Florida	N/A	25 families: 13 Caucasian families, 12 African American families; 50% single parents	Focus groups that lasted about 2 hours	Parental evaluations of treatment approaches to ADHD and the congruence of these evaluations with professional practice guidelines	Parents' accounts depicted medication as hard to accept for their children; they also said that professional guidelines are partially congruent with the role assigned to stimulants
09	2001	Claudia Malacrida	Female	Canada: Calgary and Southeast England	Foucauldian notions of knowledge, power, and resistance	34 mothers: 17 Canadian and 17 British; none were full-time employed outside home	Semi-structured interviews	Mothers' experiences with ADHD children within two different cultures	Mothers experience ADHD problem differently within two different cultures and yet similarly; phases of their experiences are described

ID	Year Pub	Name	Author's Gender	Country Where the Study Was Conducted	Theory	Respondents	Types of Methods	Aim	Findings
10	2003	Susan Dosreis, Julie Magno Zito, Daniel Safer, Karen Soeken, John Mitchell, Leslie Ellwood	Female, Female, Male, Female, Male, Female	N/A	N/A	247 parents: 219 mothers, 24 fathers, and 4 grandmothers; 92% above $25,000 per year; 75% White	Survey (47 items)	Reporting parental perceptions of stimulants for ADHD	Two-thirds of parents believed that sugar diet affects hyperactivity; 55% of parents were hesitant to use medication on the basis of information in the lay press and 38% of parents believed that too many children receive medication for ADHD. Perceptions of medication differed between White and non-White parents; non-White parents were less likely to believe in medications.

ETHICAL CONSIDERATION 3.1

RESEARCH FUNDING

While you are extracting information from articles, also pay attention to whether the researchers used an incentive or made any other statement about how they funded their research. Sometimes the source of funding may be an organization or a company that has something to gain from specific results of the study. It may be easier to collect this information while you are looking through the articles.

For example, if a pharmaceutical company is funding a study on medication adherence of patients, that company has an interest in looking at the issue from a business point of view rather than from a broader perspective. The company is not interested in knowing whether the medications are even needed for patients or the gravity of side effects. They are paying the researcher to find out about the reasons why patients are not adhering to medications so they can improve their product. These are important issues to point out in your literature review because they reveal potential biases or conflicts of interest on the side of researchers.

Using the Conceptual Graph

Once you reach this point, you probably feel quite accomplished. Let's retrace our steps. In Chapter 2, you created a graph that mapped out your constructs. There you defined your independent, dependent, controlling, and intervening variables, or even more. You have also explored the databases and handpicked peer-reviewed articles that relate to your topic of interest. You have read articles and other resources and have prepared a detailed, annotated bibliography. The information from these articles is organized into a table that will assist you in writing your literature review. In between, you have also explored various theories relating to your specific topic of inquiry.

This is the right moment to revisit the graph of constructs and variables to see if adjustments are needed. For example, the literature may have offered another layer of information or opened a door to something new worth exploring. Perhaps a new variable has emerged that was not visible in your initial graph. Often some new controlling variables may have become present that you believe are important to use in your study. The literature could have also shown that your initial topic of interest is over-researched, but that there are some unexplored threads that seem promising. It is also possible that the theory you've chosen offers a new perspective on constructs and variables that you may not have considered before.

Adjustments of all kinds can happen at this point, but once the conceptualization of variables and relationships is nearly perfect, it is time to synthesize the literature. Researchers may still want to return to specific articles for more information, but this level of organization is certain to help you avoid staring at a blank sheet of paper, not knowing where to start. Let the synthesis of the literature be a very fulfilling process!

Organizing Your Work

Most researchers have their own style of organizing the literature. Some may believe that starting broadly about a topic and then narrowing it down to their own specific niche is the appropriate path to a great literature review. Others may think that organizing the literature using various subheadings on different studies may be the key to an engaging piece. Yet others may agree that regardless of the way you organize the literature, great writing skills can make a literature review more captivating for readers. All of these approaches are great as long as the organization of the literature takes a logical path. Sometimes our thoughts can be disorganized. For example, the literature review may begin by discussing study methodologies and then jump to the theoretical approaches of the studies, followed by an idea on some dynamic that was omitted from the body of literature. This is the epitome of disorganization and it usually makes for scattered literature that does not explore any topic in much detail.

One should initially organize the literature logically. One simple way of doing so is to imagine calling a friend or parent and telling them about the literature you have read. How would you explain the body of literature and your argument in the context of the literature to someone with very little knowledge on the topic? This is one simple and effective way to understand the logic of your narrative. While it is important to organize your thoughts and put things into their own place, you should also consider writing to people who are knowledgeable about the topic. That means to give depth and logical arguments that show your engagement with the literature and your thoughts about it.

Consider all the factors that make your topic a difficult problem to investigate. Pay close attention to whether you looked at different perspectives on the same topic and whether you are making any assumptions about the problem. Recognize your own assumptions and your own perspective and state it clearly. These features will strengthen the literature review and will show to outsiders and experts alike that you have considered the problem, identified all the complex characteristics, and included various perspectives in it.

Conceptualizing Literature: Patterns

We were taught at an early age to organize things by shape, color, or pattern. You are probably an expert in organizing kitchen cabinets or messy drawers. What kind of principle do you follow? Most of us follow some form of philosophical principle that we attach ourselves to while organizing something, which we do instinctively. We do not put much thought into why we put plastic storage containers in a different drawer from the glass containers; we typically do so without thinking. Similarly, we put all the forks in a different division of the compartment cutlery box from spoons. Organizing examples are endless because they feel natural to us. The task of organizing a literature review feels like it is unnatural because we have never done it before, but the entire procedure is as simple as organizing a sock drawer.

There are bodies of literature that may be relatively smaller than other topics. One example is the literature surrounding parents of children who are diagnosed with ADHD. While there has been quite a bit of research conducted regarding other aspects of children who exhibit ADHD-like behaviors, such as adherence to medications, various types of ADHD, and similar topics, very little is known about parents' experiences when their child is diagnosed with ADHD. When the body of the literature is very small, it may make sense to organize the literature based on patterns that become visible as the researcher looks at previous studies.

In investigating these studies, a researcher will start to notice that parents of children diagnosed with ADHD seem to have some similar concerns. For example, at least four studies have demonstrated that teachers are usually the first to identify that the child has a problem (Cohen, 2006; Leslie, Plemmons, Monn, & Palinkas, 2007; Malacrida, 2001; Sax & Kautz, 2003). Other study findings have shown that parents report experiencing a trial and error stage regarding medication use for their children (Dennis, Davis, Johnson, Brooks, & Humbl, 2008). Similarly, a researcher may notice a difference in findings—something that only one study points out. These similar or different patterns could be organized into an outstanding literature review where the actual body of the literature is shaped based on similarities and differences found among these studies' findings.

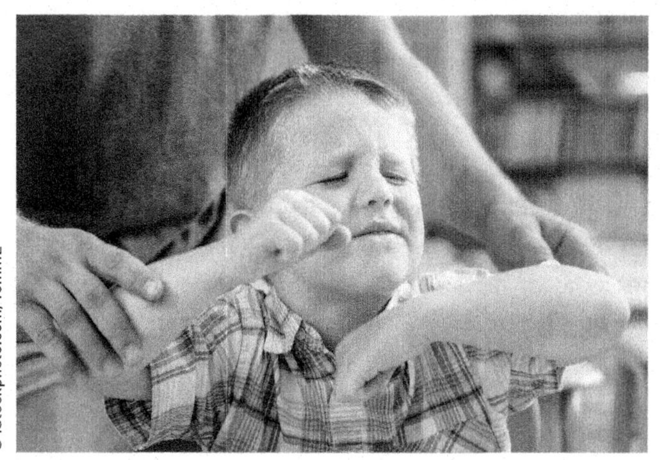

Thinking critically is the ability to entertain a variety of perspectives on the same problem.

RESEARCH IN ACTION 3.2

ILLUSTRATION OF THE ORGANIZATION OF LITERATURE

Source: Blacher, Cohen, & Azad (2014). In the eye of the beholder: Reports of autism symptoms by Anglo and Latino mothers. *Research in Autism Spectrum Disorders*, 8(12), 1648–1656.

Well-documented ethnic and racial disparities exist in the diagnosis and treatment of many physical and mental health disorders (Begeer, El Bouk, Boussaid, Terwogt, & Koot, 2009; Mandell, Listerud, Levy, & Pinto-Martin, 2002; Mandell & Novak, 2005; Smedley, Stith, & Nelson, 2002), yet recent evidence examining the prevalence of ASD among socioculturally diverse children is inconclusive. Recent studies have shown few, if any systematic differences in the prevalence of ASD among socioeconomically and ethnically diverse children (Centers for Disease Control, 2006; Mandell, Novak, & Zubritsky, 2005). . . . *(the rest of subsection is removed)*

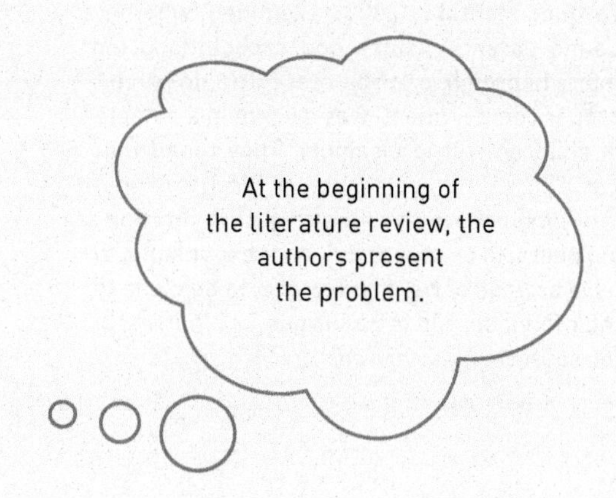

1.1. CULTURAL BELIEFS ABOUT CHILD DEVELOPMENT AND DISABILITY

Previous research has established that cultural groups vary in their beliefs about child development and disability (ref). For Latino families, research has shown that parents have distinct beliefs about what developmental skills they expect for their children and when those skills should be mastered. For example, Gannotti and colleagues, studying Puerto Rican children with disabilities and their families, found that parents expected their children to meet certain developmental milestones (e.g., drinking from a cup) much later than did Anglo parents

(Continued)

(Continued)

(Gannotti, Handwerker, Groce, & Cruz, 2001). On the other hand, Arcia, Reyes-Blanes, and Vazquez-Montilla (2000) examined whether Latino parents' values and expectations for their children (e.g., to be respectful, to exhibit proper comportment) was evident in a sample of children with a disability. They found that these parents expected their children to exhibit the same behaviors as their typically developing peers: to be respectful, and responsible, to have a sense of right and wrong, to be close to their families, and to be curious. . . . *(the rest of the subsection is removed)*

By using subheadings, the authors have divided the literature based on topics. Here, they discuss research conducted on cultural beliefs about children with disabilities.

Note how the researchers focus the discussion on Latino families because this is the ethnic group they are interested in.

1.2. AUTISM DIAGNOSIS AND LATINO CHILDREN

Evidence exists for a universal set of symptoms associated with autism, and there is reliability in the diagnosis of ASD and the differences in subtypes (American Psychological Association, 2000, 2013; Fombonne et al., 2004; Hill et al., 2001). . . .

A more recent study by Magaña and colleagues (2013) compared ADI-R scores in Latino adolescents to a matched group of non-Latino White adolescents. They found no significant differences between the two groups on

Here, they also emphasize the differences between studies: The first study found that Latino parents expect milestones later than other parents, whereas another study found that they expect children's milestones at almost the same time as Anglo parents.

the Social Reciprocity and the Communication subscale scores, but they found a consistent difference between the groups in the Restricted, Repetitive Behaviors subscale score. The Latino parents reported significantly less impairment in this area. The authors note that it is difficult to determine whether these differences were attributed to actual behavioral differences between the two groups, or whether parents may have understood the items differently (Magaña & Smith, 2013). . . .

In the next section, the authors move even closer to their own topic. Here, they explore research focused on autism among Latino children. Later in this section, the authors provide a bit more information about the nuances of a specific study that compares scores of social reciprocity and communication. They point out differences and say that it is unclear whether these differences are due to children's behaviors or parents' interpretations.

Building from previous studies illustrating the underdiagnosis of ASD among Latino children (CDC, 2006, 2014), the present study compares parents' reports of ASD symptoms with clinician observations, in a sample of newly screened children with an ASD.

At the very end of the literature review, the authors tailor the direction of the discussion toward their study and how their study simultaneously completes the literature and sheds light on some new aspect.

THINKING CRITICALLY

Exploring different venues regarding how to design your literature review is always recommended, but one important principle should be kept in mind: employing critical thinking. Using critical thinking is one of the reasons students go to school in the first place. Reading books and learning new information can help to gain knowledge. Some of that knowledge is retained and some is forgotten. Being critical, on the other hand, is a skill that can be applied to various aspects of life. But how can one be critical and what does critical thinking really mean?

Reading Critically

Critical thinking begins with critical reading of the literature. One of the best definitions of critical thinking is provided by Elder and Paul (2001), who define critical thinking as "the mode of thinking about any subject, content, or problem—in which the thinker improves the quality of his or her thinking by skillfully taking charge of the structures inherent in thinking and imposing intellectual standards upon them" (p. 19). One way of thinking about this relates to passive versus active reading. When we read something because we have to, but have no interest in the topic, we often read it passively, simply trying to retain the information we are reading. Our mind wanders to other things and although we read the letters, words, and sentences and spend valuable time doing so, we are passively absorbing information. It comes as no surprise that we may not remember much from what we are reading because our attention and focus are not quite there. To put this bluntly, this type of passive reading is a waste of time.

Parents of children who exhibit ADHD-like behaviors often find themselves in the middle of a controversy, uncertain on how to interpret their child's behaviors.

There are other times, however, when we are reading something that grabs our interest. We are reading the information in front of us, but we are also simultaneously thinking and *considering* the information given. We may or may not agree with what we are reading and for each sentence that our eyes read, our brain also completes it with additional ideas, suggestions, or illustrations. This is active reading, and it immediately leads to critical thinking because we are not simply absorbing the new information; we are agreeing or disagreeing with it and are offering our own arguments.

Critical thinking is one of those crucial skills that, if practiced enough, can become your way of reading everything. When writing a literature review, you are not simply reiterating information about the literature you read. You are raising questions, gathering relevant information, interpreting it through your perspective, and coming to logical conclusions about the material at hand. When thinking critically, you are evaluating the study you are reading, constantly checking for different perspectives, and keeping an open mind about what is missing and how the problem is investigated.

Analyzing Studies

To analyze a study, first look at the concepts, theories, and perspectives the researchers are using by asking questions regarding their intentions and whether they considered other ideas about their work. Here, you need to pay close attention to possible assumptions that authors may have taken for granted.

Next, you look closely at the methodologies employed by most peer-reviewed studies. The simplest way to remember the details of each study is to write down its methodological approaches (e.g., an in-depth interview with 35 participants of an average age of 30, a questionnaire distributed to 250 participants, a focus group with 10 participants). Often, these studies have similar methodologies and you may be able to draw a clear picture of the literature you are reviewing by pointing out the percentage of studies that used questionnaires, the number of studies that have more women participants, or other distinguishable characteristics. These details are necessary to not only portray your mastery of the literature, but also show how your study is of significance since you may be fulfilling a missing approach from these methodologies.

Let's illustrate this point. There have been quite a few studies conducted on diabetic children and what type of treatment works best or what type of difficulties these children's parents face in coping with their children's problems due to diabetes. A novice researcher may review all of the studies and write down the methodologies used. Once this has been completed, it is easy to see that many studies use multiple-choice questionnaires to survey a large number of parents. You can also see from your table that most of the parents responding to the surveys are mothers. If you were thinking of conducting in-depth interviews with both parents to delineate the daily life for a parent of a diabetic child, you are in luck. You can clearly make the point that the literature is missing an in-depth understanding of this group of parents, their daily challenges, their frustrations, their joyful moments, and their achievements. You can also point out how the majority of the participants in the studies you reviewed are mothers. Therefore, the literature is missing one crucial dimension: the perspective of fathers. This type of argument makes your literature review much stronger, more effective, and engaging. You are showing that you have knowledge of the body of literature you investigated, have found the missing gaps in it, and are contributing to the field by filling those gaps. You must look at the body of literature and find your niche. This process is simultaneously creative and exciting.

HYPOTHESES

The hypotheses of your study need to be included at the end of your literature review. After all, that is why you wrote the entire argument with the literature. You needed to prove how important your hypotheses are and why this particular study is worth investigating. Often, the alternative and null hypotheses are placed in a separate paragraph. Remember our discussion from Chapter 2 where we always try to reject the null hypothesis. In a nutshell, that is the focus of our study, so we need to state these hypotheses at the end of the literature review.

For example, I am interested in conducting a mixed-methods study about the ways that psychiatrists diagnose their patients. My goal is to find out the process of how a person is diagnosed with any disorder or mental problem, as well as the percentage of people who walk in the door of a psychiatrist's office, but are mentally healthy and leave without a diagnosis. Once the literature is explored and it shows that there is a disagreement among researchers on whether some diagnoses are truly mental problems as well as the increasing number of people being diagnosed in our society, the literature review clearly shows the need to explore this topic further. At the end of it are the hypotheses. The following is an illustration of this example:

H_1: There is a relationship between the number of complaints people express at the psychiatrist's office when they walk in for the first time and whether they leave with a diagnosis. The relationship is one-directional, so the higher the number of complaints, the higher the number of diagnoses the patient receives.

H_{01}: There is no relationship or trend between the number of complaints from people who walk in to a psychiatrist's office for the first time and the number of diagnoses they receive.

H_2: There is a relationship between the number of complaints people express at the psychiatrist's office when they walk in for the first time and whether they leave with a prescription for psychotropic medications. The relationship is one-directional, so the higher the number of complaints, the higher the number of prescriptions.

H_{02}: There is no relationship between the number of complaints of people who walk in to a psychiatrist's office for the first time and the number of prescriptions they receive.

SYSTEMATIC REVIEWS OF LITERATURE

A **systematic review** of the literature is different from a literature review. In fact, a good systematic review of literature has its own literature review. A systematic review is a form of scientific study. It is an attempt to bring together every scientific study on a specific topic and draw conclusions from it. A literature review tries to prove why the researcher's topic is a topic that needs to be investigated and it is tailored to support the researcher's study. A systematic review has no such agenda. It is focused on synthesizing all the literature on a topic. Let us look at the differences between a systematic review and a literature review for a better understanding.

Systematic Reviews Versus Literature Reviews

There are four major differences between a systematic review and a literature review, which we will discuss in detail here.

(a) *The focus is different.*

When we conduct a study on a topic that interests us, we are prone to have collected studies that support our ideas or even read scientific reports we find appealing. Sometimes we may even disregard or not pay close attention to a study that investigates something completely different from what we are trying to accomplish—and we do this unintentionally—mostly because we are truly dedicated to our topic. This translates, however, into a literature review focused to serve the study we are undertaking rather than a review of literature for the sake of review. A systematic review focuses on systematically collecting together every scientific study in a topic of interest and evaluating them. The focus of the systematic review means conducting the best review possible rather than using it to serve another study. This basic differentiation between these two forms of reviews is at the root of the rest of the distinctions between the two.

(b) *Selection bias is reduced.*

A systematic review follows specific guidelines in collecting studies on a topic. This means systematic reviews are less prone to biases, such as **selection bias**, that may result from choosing different studies as is done in traditional literature reviews. Selection bias refers to the researcher's tendency to look closely at scientific articles or scientific work that aligns with that researcher's ideas about a topic and overlooks other work that may oppose these ideas. The lack of biases and thoroughness of systematic reports are what makes these pieces of scientific reports very attractive. A systematic review reports exactly how many databases were searched, the rationale of choosing those specific databases, what keywords or combination of keywords were used, and how many results each search yielded. They are inclusive of all the studies that fit the predefined criteria of searching. The researcher cannot decide against including a study in the review without a specific reason that needs to be reported.

Traditional literature reviews are not as strict, although it is widely believed that most researchers follow exactly the same steps for traditional reviews and that a strong traditional review has some elements of the systematic review incorporated in it. However, since criteria for traditional reviews change, there exists the possibility for potential selection biases.

(c) *There are rigorous steps and procedures.*

A systematic review of literature is a research study that focuses on other scientific studies. To put things in context, the scientific studies that are examined from a systematic review of literature are the participants of the study. As such, conducting a systematic review implies following rigorous steps and procedures to achieve the best quality possible. When we extract scientific studies for traditional literature reviews that are supporting specific research, we go to various databases, try different keywords, look at studies that we find from the references of a major study, and so on. We rarely keep a strict record of how we found out about these studies and what databases we used. These steps and procedures on how to extract scientific studies on a topic *are* extremely important for a systematic review of literature. This is the **methodology** of the systematic review because it answers the question on how the data were collected—in this case how the studies were collected—what sources were used and why, how the researcher made the decision to include specific articles, but leave out some others. In other words, the methodology of systematic reviews is the methodology used to establish that the collection of articles in a topic is sufficiently inclusive, unbiased, systematic, and following methodological guidelines.

(d) *Assessment and analyses of studies are involved.*

A traditional literature review will often provide some information on one or two studies and will summarize the conclusions of another study, depending on how the researcher has designed the review. A systematic review of literature will collect all the methodologies of all the studies, conclusions, and discussions and will provide summaries as well as analyses of these scientific studies. In other words, it will thoroughly investigate the data from all the studies and provide detailed information about these studies and their similarities and differences.

Meta-analyses and systematic reviews of literature are very similar, with one important difference. Meta-analyses statistically recalculate the data from the original studies based on standard criteria. The procedures for these calculations are beyond the scope of this book.

RESEARCH WORKSHOP 3.1

WRITING THE LITERATURE REVIEW

There are different ways of organizing your literature review, but one simple way to get you started may initiate from the table where all your literature is mapped out along with your annotated bibliography. Taking these simple steps may give you some ideas on how to write a solid literature review:

1. Look out for things that are similar to or different from all (or most) of the articles. Anything from the types of methodologies used to participants, use of theory, or findings. Sometimes the same theory is overused and you may be thinking that a different perspective may help this same topic. Other times all the participants may be from the same gender or the same ethnicity, and you may make the point that we need more diversity on the issue.
2. Write down three or four points that you notice that are either the same across all studies or different (if different, group them).
3. Organize your literature in paragraphs, first by thinking broadly to then a narrow focus—this focus is the argument why your study will be an important addition to the current literature and different from it.
4. Write each point in its own paragraph with very simple sentences first and then expand by adding information you may have already written in your annotated bibliography.
5. The very end of your literature review should have some argument why your study is important to the scientific knowledge in this topic and how it will be different.
6. You may end by stating your own research question or hypotheses—but those can also be in their own section right after the literature review.

Summary

Writing a literature review is a rewarding task with added perks, such as (1) painting a picture on what the scientific literature in your topic really looks like; (2) pulling out studies that are relevant to your topic and showing your organizational skills in making sense of them; (3) displaying your creativity in designing the literature review that best supports and complements your study; (4) exercising critical thinking by looking at the studies from various perspectives; and (5) writing a logical, critical, and analytical literature review that supports your proposed study and naturally ends with your research question(s) or hypotheses.

In this chapter, you were introduced to some helpful tips on how to organize the literature and, most importantly, how to initially conceptualize the tasks. This organization is unique to each researcher, but some helpful tips were provided in how to look for patterns, similarities in methodologies or findings, as well as distinctive differences or opposing results from studies. Your literature review will culminate with the hypotheses or research questions you have formulated. For every alternative hypothesis, we include a null hypothesis because the null hypothesis is the one we will eventually try to reject.

This chapter also introduced systematic reviews of literature and how they are different from the traditional literature reviews. A systematic review of literature tends to be protected from selection biases—biases related to how we choose studies to include in our review. A systematic review also follows rigorous steps and procedures compared to the traditional review. After all, a systematic review of literature is best conceptualized as a research design rather than a literature review. It usually has its own traditional literature review. Systematic reviews are also more organized and full of details on the studies they have investigated. This allows researchers to paint a complete picture on the literature of a specific topic. Meta-analyses are similar to systematic reviews with an added feature of statistically recalculating the data from original studies.

Key Terms

Annotated bibliography: a brief summary of the article or book you read, including the focus of the study, the methodology used, the findings, and any other important information that directly relates to your research topic.

Boolean operators: operators that are used to conduct searches in the library and other databases (i.e., AND, OR, NOT).

Interlibrary loan: the possibility of borrowing articles and books from other libraries that may not be available in your local or university library.

Literature review: the body of literature surrounding a specific topic of interest to the researcher.

Meta-analyses: the type of systematic reviews that statistically recalculate the data from the original studies based on standard criteria. Meta-analyses are a rigorous way of conducting systematic reviews.

Methodology of the systematic reviews: answers the question about how the articles and books (and other media) included in the systematic review were collected, what keywords were used, what techniques were used to narrow down the number of published records, and what rules were followed for selecting the final sample of articles and books.

Selection bias: biases that surround the way we select and decide which articles to use for our literature review. To avoid selection bias, we often conduct a systematic review of the literature on the topic.

Systematic review: a review of the literature that follows specific guidelines in collecting studies on a topic. It is usually less prone to biases.

Taking a Step Further

1. What type of information is commonly included when writing an annotated bibliography? (Name all that apply.)
2. How is the systematic review different from a literature review?
3. What distinguishes meta-analyses from all other types of reviews?
4. What is the role of hypotheses on our literature review?
5. How can we start our literature review?
6. What is *selection bias* and what types of reviews are more prone to include selection bias?

4

QUANTITATIVE DESIGNS

CHAPTER OUTLINE

WHAT WILL YOU LEARN TO DO?

1. Describe the purpose of exploratory, descriptive, and explanatory studies
2. Compare and contrast cross-sectional and longitudinal studies
3. Explain the differences between nomothetic research and idiographic research
4. Discuss each type of experimental design and its advantages and disadvantages

CATEGORIZATIONS OF RESEARCH STUDIES

The next step in our study involves designing our proposed work. How will we do this? What type of tool will we use to gather the information or data we need? Where will we find participants? The answers to these questions and others will become clear in the design process. This chapter examines quantitative designs while the next chapter will delineate qualitative designs. One simple way of conceptualizing the design of our study is to ask a few questions. The answers to these questions will help us with some of the basic characteristics of our study and make the entire process much easier. After all, designs are a combination of two or more of these characteristics or identifiers.

Designing a study is similar to designing an empty room. There are some major things that you need to decide in order to begin—a few characteristics that are crucial before you pick up a brush and start painting the walls or placing furniture in the room. First, you need to determine the room's purpose. Is it going to be a kitchen, living room, playroom, or bedroom? That is exactly where you would start with research design as well—conceptualizing the purpose of your study.

Then, you may consider some basic details about the room, such as paint color or the type of flooring. The same process applies for research—once the purpose of the study is clear, you will consider the basic details relating to participants and how to contact them.

Once these features are set, you will choose your furniture, determine where those pieces will be placed, and consider other details, such as an approximate budget and unexpected surprises. Designing research is not so far removed from this description. As researchers, we need to visualize the study, determine the practical details of the study, learn about the possible errors that could occur, and finally, begin working. Let us examine some characteristics of research design that researchers must consider when shaping their study.

The first characteristic of research design involves *what* we are trying to accomplish in this study. Are we trying to *explore* a new relationship between variables? Are we trying to *describe* some phenomenon in detail? Are we trying to *explain* an established relationship between two or more variables? The most common distinctions of studies are based on whether we are trying to explore, explain, or describe something. That marks the first categorization of studies.

Exploratory Studies—Answering "What?"

Exploratory studies are often used to investigate a new topic of research, a new thread of previously established relationships, a new methodology, a new instrument of data collection, or to gain a deeper understanding on a specific population. These types of studies are useful as initiators for larger, more complex studies and are often thought to pave the way for sturdier, more sophisticated research. Exploratory designs refer to studies that attempt to explore unknown territories. They aim at investigating, exploring, or attempting to figure out a new, innovative thread of knowledge. Exploratory studies can be quantitative or qualitative.

Sometimes, researchers may have a hunch about something, but they cannot find sufficient literature to support their idea. In such cases, it makes sense to undertake an exploratory study that may confirm or refute this initial hunch. Exploratory studies are less expensive than larger studies, which may explain why they are popular and easy to conduct. To decide whether a study is exploratory or not, we need to consult the literature review. Is your specific topic an innovation that is not even included in the literature? An exploratory study usually explores a new relationship between variables or an existing relationship between variables, but on a new population.

For example, say a researcher is attempting to explore the relationship between stigma and mental health diagnoses among adolescents. If the relationship between one's perceived stigma and that person's likelihood to seek mental health treatment is not analyzed in any other study, this is an innovative, exploratory study. If the relationship was already established by other researchers, but none of them have examined adolescents as a population of interest, this study still remains exploratory research. In other words, when either the relationship between variables (stigma of mental health and seeking mental health treatment in this case) or the population has not been previously established from the literature, the study is exploratory.

Another kind of exploratory study may occur when the study combines different pieces of established relationships together. Entertaining this example further, if the current literature has established that (1) there is a relationship between the stigma of mental health and seeking treatment for mental problems and (2) there is a relationship between seeking treatment and suicidal thoughts/actions, we may be interested in exploring the relationship between the stigma of mental health and suicidal thoughts/actions. The relationship may not have been explored before even though pieces of it have been found in the literature.

Descriptive Studies—Answering "How?"

As a young researcher, you may be more interested in how a phenomenon happens than in attempting to establish a relationship between two or more variables. For example, we know that there are illegal immigrants in the United States—the statistics speak for millions of such people around us. We know that many of these people left their country in search of a better life for themselves and their children, to escape problematic circumstances in their homeland, or even because they perceive moving to another country as upward mobility. Regardless of the situations that caused these people to move, a researcher may be interested in knowing how they survive in the United States illegally. How do they find work? Where do they work? What is their day like? What are the practical and emotional difficulties they and their children face? What does it mean to be stigmatized? Do they face judgments and prejudice every day?

Descriptive studies are generally qualitative because it is often necessary to gather in-depth data from people and allow participants the flexibility of expressing their feelings and situations in detail. Descriptive designs allow researchers to focus on describing a phenomenon or understanding the details about people's experiences of a particular event. In the case of our immigrant study, a descriptive design would allow the researcher to advance the knowledge on this group by depicting a more detailed picture of how their situation develops. On the surface, one may wonder about the purpose of better understanding people in a specific situation. But if we are aware of the difficulties that a specific group of people is experiencing, we can think about possible ways of intervening to help the group in need.

Many social workers are interested in the end-of-life process, for example. It has taken many descriptive studies to understand how people feel when they become aware of death. Other descriptive research has focused on how friends and family members face the situation of preparing to lose a loved one. These important details from research have improved and refined the approaches taken by hospices all over the country. Without descriptive studies, we would not be able to really understand the experiences, the difficulties, or the emotional toll that a particular situation takes on people. Further, we would not be able to aid them in any way if we are not aware of what they are going through.

Explanatory Studies—Answering "Why?"

Explanatory studies are reinforcing studies. They focus on explaining the reasons behind a phenomenon, relationship, or event. Explanatory studies are very helpful in understanding the question "why?" and clarify the degree of influence of one variable versus another on the dependent variable. This type of research does not set out to explore something new, nor does it attempt to gain a deeper understanding of how something works. Instead, explanatory studies seek to explain the reasons behind an already known and/or established relationship.

For example, there may be an established relationship between family substance abuse and future illicit drug use of children raised in these families. An explanatory study may attempt to further explain the relationship by exploring why a child raised in this environment is at risk for illicit drug use. Measuring a number of other variables, such as socioeconomic situation, depression and/or other mental health concerns, marital status, employment, education, and other characteristics among individuals who were raised in substance abusing families can help the researcher make additional connections in the already established relationship.

Explanatory studies can be quantitative or qualitative—it depends on what the researcher decides is better for the specific study. For example, a researcher may be interested in understanding the reasons why some victims of domestic abuse stay in their relationships. Oftentimes, these women never report their abuser and may never leave the relationship. The reasons these women stay in their relationships can be investigated either quantitatively or qualitatively, as each approach has different benefits. An explanatory qualitative design may bring forth some reasons that the researcher had not considered. An explanatory quantitative design may list the possible reasons and ask participants about them to determine which reason carries the most weight and keeps the women in their abusive relationships.

CROSS-SECTIONAL VERSUS LONGITUDINAL STUDIES

Once a study's purpose is established, the researcher knows whether he or she plans to explore, describe, or explain something. Just like the empty room example, the designer knows whether the empty room will be a kitchen, living room, playroom, bedroom, or bathroom. Now it is time to consider who will use this room—the entire family, the children, or the parents? Similarly, we need to examine another layer of characteristics: the *frequency* of data collection. How will we collect these data? How often will we meet with participants to collect data? Once? Twice? More? The frequency of contact divides studies into two types: cross-sectional and longitudinal.

© iStockphoto.com/Rawpixel

Are we collecting data just once from the participants, or are we meeting with them a number of times and collecting information every time we meet with them?

Cross-Sectional Studies

Cross-sectional is the name for studies that collect data only once. It is a simple snapshot of data collected at one point in time. For example, conducting a study that collects information from participants who have purchased an expensive car in the last year. Collecting cross-sectional data refers to the fact that these participants were approached just one time to collect the data. Now, if these participants were approached a few times and data were collected each of these times, the study is not cross-sectional anymore. Cross-sectional truly refers to one shot of data collection, without any follow-up questions. These types of studies are useful when we are interested in collecting data from a large number of participants. Because of time and cost constraints, researchers are attracted to cross-sectional designs. It should be noted here, however, that cross-sectional studies may collect the data over a specific period of time (like weeks or months because of the difficulties relating to data collection), but the data are only collected once from each participant. Researchers cannot follow up with participants or ask for their opinion again to see if their answers have changed.

Cross-sectional designs are widely used because of their convenience. The data are collected relatively quickly, which may mean fewer expenses and a shorter time frame, which are very attractive features for researchers. Regardless of their attractiveness, there are some disadvantages to this type of design. It is difficult to establish any cause-effect relationships when using cross-sectional design because all the variables are collected at the same time. This means that it becomes very difficult to state whether a specific feature was always present or was caused from something else.

For example, say a researcher is interested in investigating the relationship between anxiety attacks and illegal substance use among teenagers. Because of time and cost constraints, this researcher decided to use a cross-sectional design to collect the data. Further, a few trained interviewers were hired to collect the data by asking high school students a number of questions. Four hundred students from an urban area took part in the study and the data took a month to collect. The participants were asked about their illegal substance use, their anxiety attacks, their demographics, and family ties.

In analyzing the data, the researcher found an impressive correlation between illegal substance use, anxiety attacks, and poor family ties among these teenagers. The results were intriguing, but there was an important piece missing. Did the anxiety attacks trigger the illegal substance use as a form of self-medication, or did the substance use cause a chemical imbalance that led these teenagers to have anxiety attacks? Furthermore, did poor family ties create the substance use problem and the anxiety attacks, or did the drug use and anxiety attacks prove detrimental for family ties? This is an important aspect that is often lost when the data are collected all at once and these aspects are present. Then it is impossible to say whether one variable changes another or the other way around.

Another problematic feature of this design has to do with the fact that people are often volatile and their mood may influence how they answer the questions. Sometimes we talk to them at the wrong time, and because of other extraneous circumstances, we may end up with incorrect information. Sometimes participants may be feeling unhappy, hungry, or lonely or are having a bad day when we are trying to collect our data, so their emotions and feelings may influence their responses to our questions. On a regular or different day, their answers may change completely. This type of error will likely be present in our cross-sectional study, and we can only hope that it is minimal and that the large amount of data we are collecting reduces this possibility.

RESEARCH WORKSHOP 4.1

THE ADVANTAGES AND DISADVANTAGES OF CROSS-SECTIONAL DESIGNS

Cross-sectional design studies are a great tool to collect a lot of information in a short period of time and researchers should be as thorough in their measurements as they can. This type of design has a number of advantages and disadvantages to consider when selecting your study's design. Most importantly, the design should serve the hypothesis or the research question, not the other way around. The following is a brief list of advantages and disadvantages of cross-sectional designs to help you make the right decision.

ADVANTAGES:

(a) Cost: inexpensive

(b) Time: takes a relatively short time to collect data

(c) Spread: collects information about many variables

(d) Relationships: allows for investigating new relationships between variables

DISADVANTAGES:

(a) Lack of causality: because data are collected at one time, there is no possibility to prove cause-effect relationships

(b) No turning back: if there was a mistake in the data collection, there is no way to fix it because there is no follow-up of any kind

(c) Vulnerable to disturbing variables: some extraneous and unpredictable variables can create bias in your results and you may not even be aware

Longitudinal Studies

Longitudinal studies collect data at different points in time, which helps the data to be more accurate and avoid or minimize errors like inaccurate responses. These different data collection points are called *waves*. The first data collection is often called *base wave* or *wave I*, and usually every data collection point after this is named *wave II*, *wave III*, and so on. Known for their accuracy, longitudinal studies are quite desirable for various topics.

For example, we may want to know the long-term effects of a specific medication given to children. Although it seems that the medicine works fine in the short term, we cannot be certain about its long-term effects if it is used daily from an early age. In cases like this, longitudinal studies are not only desirable, but are the only way to find out the outcome of a specific medicine or other treatment plan. In this type of study, researchers may follow a number of participants over the years and collect their data routinely at specific times. Alternatively, they may decide to gather information from young children who are taking the medicine, slightly older children with a history of taking the same medicine, and even a third group of young adults who have been taking this same medication.

One major disadvantage of longitudinal studies is the fact that they require quite a bit of time to collect the data—depending on the study, this could take years. Because of the length of time, these types of studies are more costly. There are more people involved in gathering, cleaning, analyzing, and interpreting the data. Depending on the nature of the study, it may also entail various uses of technology, incentives for participants, and other costs. The following types of longitudinal studies are widely used in research: panel studies, trend studies, and cohort studies. Let us take a glimpse at these different types.

ETHICAL CONSIDERATION 4.1

INFORMED CONSENT DURING A LONGITUDINAL STUDY

A longitudinal study can last for many years. It is not uncommon for a participant to be an infant when the study begins and continue participating until late adulthood. The researcher received initial informed consent from a child's guardian or from the adults who are participating in the study. However, this initial informed consent does not imply continuous informed consent for years even if participants are aware that the study will last for a long time. During the second or third wave, participants may change their minds or if a participant was a child at the beginning of the study, may not want to participate as an adolescent.

During each wave of data collection, participants need to renew their informed consent and be reintroduced to the aims of the study. They should be told or reminded that their participation is voluntary and they may drop out of the study at any time.

Panel Studies

Panel studies are longitudinal studies that follow and accumulate data from the same participants over a period of time. The group of participants is called a panel. These types of studies are useful in thoroughly understanding topics like aging, the development of a disease, and developmental changes in career. The advantage of following the same group of participants relates to the fact that all possible environmental, biological, psychological, or social aspects are taken into consideration. In this way, many possible errors are minimized.

A great example of a panel study is a national study of socioeconomic status and health—the longest longitudinal household survey in the world. It is called the Panel Study of Income Dynamics (PSID): a national study of socioeconomics and health over lifetime and across generations. The study began in 1968 and is still running; data collection occurs yearly. It has gathered information from 18,000 individuals in 5,000 U.S. families. This study is directed by the faculty at the University of Michigan and includes a wide variety of information on employment, health, wealth, expenditures, marriage, childbearing, education, and many other topics. The data are collected from the same 5,000 families and their family members every year. This has created a large dataset with rich information on many topics that has allowed many researchers to analyze, understand, interpret, and draw conclusions on various topics of interest.

Though they are known for their accuracy and are quite desirable, panel studies are costly and are often limited in size. This type of study suffers from another drawback: sample size attrition. This means that participants drop out prematurely and do not participate in the second or the third wave of the study. There are many reasons for attrition: Sometimes people move out of the country and are unavailable, they pass away, or they are simply unwilling to continue giving their information to the researcher. The University of Michigan study described is so large that attrition may not have had an impact, but many panel studies are smaller.

Trend Studies

A **trend study** is a type of longitudinal study that collects data at different points in time from different participants of the same population. Its main focus is capturing changes or *trends* of a specific population on a particular issue or concern. The participants relate to each other because they have experienced, been part of, witnessed, or have opinions about the same event or trend. In research terms, this means that these people share one specific variable that is of interest to the study. For example, a trend study may focus on American mothers' experiences, feelings, perceptions, and practices regarding breastfeeding over time. Has there been a trend regarding how American mothers perceive

There is so much information about breastfeeding or formula feeding that can easily overwhelm new mothers.

and practice breastfeeding over time? The participants are all part of the same population: American mothers—that is the constant variable that they share.

To determine changes in this trend, the researcher needs to collect data from American mothers in 1990, and then another wave of data from other American mothers in 1995, then 2000, 2005, 2010, and 2015. Once all the data are processed and analyzed, the researcher may notice changes in attitudes, perceptions, and practices from mothers regarding breastfeeding. The trend may show that American mothers in the early 1990s believed that breastfeeding was an inferior way of feeding infants compared to formula. They also perceived breastfeeding as a hindrance on their working careers and their personal freedom and deteriorated their bodies.

Based on the data, this hypothetical study finds that in 1990 and 1995, American mothers avoided the practice of breastfeeding. However, in 2000 and 2005, American mothers began to shift their perceptions about breastfeeding. They started believing and perceiving breastfeeding as the best nutrition for their babies, and felt that formula was an inferior substitute for breast milk. During this period of time, their perceptions also changed regarding what breastfeeding did to their bodies; rather than viewing the practice as harmful to their health, there was a shift in this trend toward believing that breastfeeding had health benefits for mothers' bodies as well.

As the data accumulate over the years, the researchers see a third change in the trend. In 2010 and 2015, mothers seem to have come to a better balance in their perceptions, attitudes, and practices regarding breastfeeding. In this timespan, it seems that most American mothers have no real hierarchical order when it comes to what type of food is more nutritious or best for their babies. They feel that both practices are equally wonderful for infants and mothers will do whatever is best in their own circumstances.

As you can see from our hypothetical study, trend studies are very useful in comparison studies or studies that focus on specific changes of the population. A trend study may be able to determine many different phenomena, like the variety of practices from breastfeeding mothers around the world, how aging is approached from various cultures, or changes in childrearing practices.

Cohort Studies

A third type of longitudinal study is a **cohort study**. These types of studies follow the same cohort of people over time. You are likely familiar with the word *cohort*. All the students who started college at the same time as you did are considered to be your cohort. Students who will graduate with you from the same major are also in your cohort. This does not mean that they share your age, gender, ethnicity, race, sexual orientation, family income, marital status, or any other characteristics. The one characteristic you share is graduating at the same time or starting school at the same time.

In research, a cohort includes a broader definition, albeit similar to the one we use in universities. A cohort may include participants who were born in the same year, were married in a specific year, had their first child at a certain time, or bought their first car at the same time. A cohort study follows people in the same cohort for a designated period of time. Unlike panel studies, these are not the same people who are interviewed over and over at different points in time. Unlike trend studies, these people do not share a characteristic or a variable at the time they participate in the study. These are different people from a specific cohort of the population who have done something of interest to the researcher at the same time.

Imagine that a researcher is focused on understanding how children learn and how that learning changes as they grow. This researcher is attempting to conduct a study that follows a cohort of children born in 2000. That is the population or cohort of interest. He selects a sample of these children in 2005, 2007, 2009, 2011, 2013, and 2015. Each of these six waves of data collection includes a different group of children who participate in the study, but all of them were born in 2000. After analyzing the data, the researcher notices that as the children grow, their attention to detail is decreasing. Their interest in learning new things was very high when they were 5 and 7 years old, but has decreased as they reach adolescence and turn 13.

CAUSALITY IN RESEARCH

An important feature of design that requires special attention is determining the type of relationship our study will investigate. Once we know (1) the type of study we are conducting (i.e., exploratory, descriptive, or explanatory purpose) and (2) the frequency of data collection (cross-sectional or longitudinal), we may be wondering about the relationship between variables that we will investigate. Researchers often attempt to establish a causal relationship between variables of interest. Casual relationship between variables refers to the researchers' attempts to determine that one or more variables (the independent variables) have caused the changes in another variable (the dependent variable). We may be hoping to find out the causes of truancy among teenagers or the effects of parents' substance abuse on children.

Nomothetic Research

Regardless of our topic of interest, we may shift our attention to the type of causal relationship we are interested in. Are we attempting to test a causal relationship between the independent variable and the dependent variable that can be generalized to the entire population or are we trying to explore particular causal relationships for one participant at a time? This is the philosophical division between nomothetic and idiographic research. **Nomothetic research** aims at arriving at causal relationships that are easily applied to the population at large, such as establishing that people with a higher education level will likely earn a higher salary.

Nomothetic research is often quantitative because of its large scope. For example, a researcher may be interested in testing how our moods influence our willingness to donate to charity. Once the data are collected, the researcher establishes a causal relationship between feeling very good about oneself and a higher willingness to donate. This type of causal relationship attempts to establish general laws regarding how most people behave when they are in a positive mood.

To establish nomothetic causality, a research study needs to establish three criteria: (1) **correlation**, (2) **time order**, and (3) **non-spuriousness**. Correlation means that the variables we are studying are related to each other. In other words, the changes in our independent variable should go hand in hand with changes in the dependent variable. Correlation establishes an association, some form of obvious systematic relationship between variables, but it does not prove that the independent variable is causing the changes in the dependent variable.

Let's consider a simple example in our daily life. You may notice that when you have money in your pocket, you spend it. The more money you have available, the more you spend. It almost seems that the *need* to buy something relates to how much money you have available. Is the need causing you to carry more money or is having more money increasing the need to spend it? The need and the amount of money are definitely correlated with each other, but this does not mean that the direction of causality is determined. Correlation is simply an association between the independent and the dependent variable, but it is not sufficient to prove causality.

To be able to prove causality, we need to look at the second criterion: time order. Time order determines the actual order of changes from one variable to another. If the changes in the independent variable happen first in time and are accompanied by changes in the dependent variable happening later in time, we can say that the criterion of time order is fulfilled, and it is clear that the changes in the dependent variable occur after those of the independent variable.

In your daily life, you do not have to think of rules that will prove that the new laundry detergent you are using is much better than the old one. You know this is the case, because you bought a new detergent and tried it on a load of clothes and the outcome of the clean clothes was much better. In other words, you found a correlation between clean clothes and your new detergent. The time order is important for confidence in your findings—you used the new detergent and then your clothes were cleaner.

What if you accidentally took a bin of clean clothes, thinking that they were dirty, and washed them again? Since they were washed twice, these clothes are naturally cleaner than before, but that has nothing to do with the new detergent. The relationship between the detergent and cleaner clothes is *spurious*. A spurious relationship is more than just a false one. It is a relationship that is explained away by a third variable that you may not have thought about. In the daily life example, it was by accident. In order for us to be able to claim nomothetic causality, we need to have a relationship that is non-spurious, which is the third criterion of being able to claim causality.

Idiographic Research

Rather than applying causal relationships to the general population, some researchers may be more interested in exploring causal relationships in depth as they apply to a particular individual. These types of studies are called **idiographic research**. Idiographic research attempts to explain the entire realm of influences and every possible detail that has influenced a particular phenomenon or event for an individual. It is often qualitative in nature and is conducted using case studies, unstructured interviews, and other methods.

For example, imagine that a researcher is exploring the causes behind postpartum depression by looking at a handful of mothers and exploring in very fine detail their narratives and individual situations. The literature suggests that a large number of women experience some form of postpartum depression after they give birth. Conducting idiographic research allows the researcher to gather sufficient information and details about how postpartum depression manifested in each mother. After the dynamics of that particular mother are understood, the researcher may study another mother and so on. Idiographic research is time-consuming because it can investigate only one individual or event at a time because of its depth in truly capturing the information. Once the researcher has collected information from a few cases, he or she may be able to see commonalities across the mothers' situations. Maybe these particular mothers had unplanned pregnancies and are simultaneously burdened by the difficulties of raising a child and the radical change happening in their lives.

Idiographic research is rich in details and fuels later nomothetic research, but it is an important tool used to investigate a problem at an individual level before making generalizations about society at large.

EXPERIMENTAL DESIGNS

At this point, you have made many decisions about your study's design. Just like designing an empty room, you chose a topic and had very little idea about how the study would look when it came together. First, you thought about its purpose—whether it would describe how an event or characteristic happens, explore something new, or explain a phenomenon or event. Second, you considered how you would carry out the research and thought about the frequency of contacting the participants. You decided between a cross-sectional design and longitudinal design. With a longitudinal design, you also determined if you would study participants from the same cohort or same panel of people or if you would study the same trend. Finally, you decided between nomothetic and idiographic research—whether you would study general causal relationships that can apply to all members of a society or focus on individual cases.

These separate pieces of the puzzle are completing the larger picture of your research design. Your decisions have helped establish exactly what you are doing and you have put some thought into why you have chosen one aspect of your design over another, even if the answer relates to convenience or lack of time. Being able to explain why you are following a specific design and how the chosen design best fits your study makes you a researcher who is well aware about your own investigation. We have reached the moment of choosing one final design characteristic that will complete the puzzle. You must decide if your study will follow an experimental or non-experimental design. An **experimental design** is a study that includes an experiment and then analyzes the results. To conduct an experimental design, the researcher follows a number of specific guidelines that are best described in the *classic experimental design.*

Classic Experimental Design

What is an experiment? We refer to an experiment when we are trying to test something and we are not certain about the outcome, right? Toddlers do this all the time in order to understand the world around them. Most toddlers will try to experiment and touch the flickering light of the candle on their first or second birthday. It burns their fingers a little and they learn not to touch it again. Touching, smelling, and putting things in their mouth is how a toddler understands what actions may be repeated and what may not be such a great idea to do again. Researchers conduct experiments all the time, and they too reach conclusions about whether the experiment is worth sharing with the world so everyone can do it or whether the experiment is not truly beneficial to anyone and need not be repeated.

What exactly is an experimental design? It refers to the presence of a test, experiment, intervention, or some form of manipulation happening at the time of data collection. For instance, we may want to see how people's interest in a particular topic will increase or decrease if they are offered a pamphlet with some information on that topic. Maybe we want to see if people's sensitivity about a topic will increase if they watch a romantic movie. Maybe you are interested in finding out how helpful a social worker's intervention can be for inmates who are being released from prison.

One example of an experimental design comes from Amy Cuddy, a professor at Harvard University who has conducted experiments regarding how bodily power poses affect the ways that people perceive us and, more importantly, how we perceive ourselves. She discovered that people who engaged in power poses for 2 minutes a day increased their testosterone levels and their feelings of being powerful, and decreased their cortisol levels, which related to their stress levels.

In order to conduct an experimental design (see Figure 4.1), we need to fulfill three requirements: (1) the presence of a control group and an experimental group, (2) randomly assigning participants to the control and experimental groups, and (3) collecting data before and after the experiment, also known as pre- and posttesting.

FIGURE 4.1 ■ Classic Experimental Design

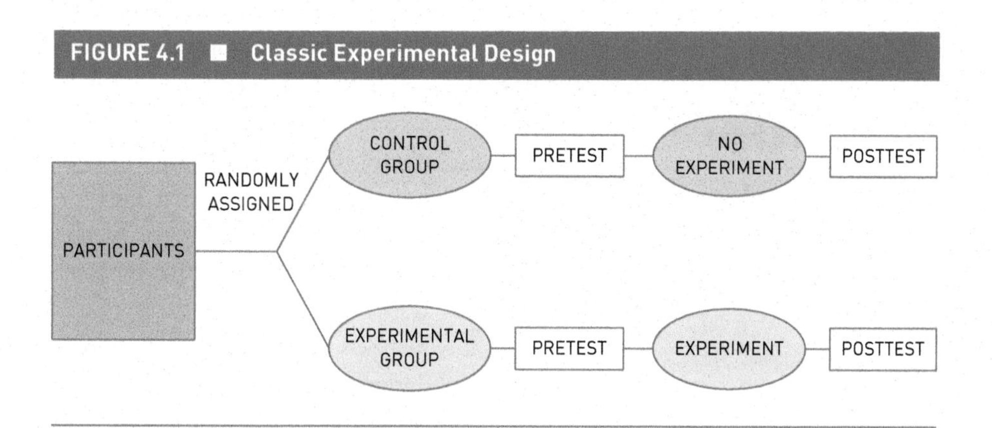

Experimental and Control Groups

Now, let's look at the criteria closely. The first criterion requires us to have an **experimental group** and a **control group** present when we collect information from participants. The experimental group is the group of participants who are undergoing some form of experimentation, such as training, taking a test or a drug, or some other form of intervention. The control group is a group of participants who do everything the participants in the experimental group do, but are not given any test, drug, intervention, or manipulation, although they may be told that they are. This group controls for the change in results of the experimental group.

The control group controls for confounding or disturbing variables. As you remember from Chapter 2, confounding or disturbing variables can affect the independent variable and, as an outcome, the results of our dependent variable. Disturbing or extraneous variables are more difficult to notice because they are not directly related to the independent variable, but are powerful enough to create problems with the results of our study. For these reasons, the presence of a control group allows us to have confidence that our findings are accurate and the outcome was a consequence of the experiment that we conducted rather than other extraneous reasons.

Random Assignment

When we are ready to collect our data, the first thing we should do is to randomly place our participants into the experimental group and the control group. Say, for instance, we have 20 participants who volunteered for our study. We would randomly divide them into two groups with 10 participants each: Group A and Group B. Group A will be the experimental group and Group B will be the control group.

For the purpose of illustration, let's imagine that we are testing whether violent video games are more likely to negatively influence people's behavior and make people less sympathetic toward various issues. Each participant in the experimental group is given a violent video game to play for about 30 minutes. After the game is over, we show them the story of a homeless family who lost everything during the downturn of the economy and ask for a small contribution.

Can violent video games influence our sensitivity and sympathy toward various social issues?

To make sure that our findings are real and because of the violent video game, our control group is given random types of video games to play for 30 minutes rather than violent ones. After playing the video game, members of the control group are also shown the case of the homeless family and asked to contribute. If our findings show that the control group was more sympathetic and willing to contribute to this family compared to the experimental group, we have been able to prove that playing video games can make a person less sensitive and less willing to give to others.

However, what if all the participants in our experimental group were teenage boys and all the participants in our control group were younger girls? Then the results of the effects of the violent video game playing may be due to other characteristics, such as gender and age rather than our own experiment. This brings us back to the second requirement of an experimental design: random assignment. Note that we randomly assigned people to the experimental and control groups. If the process of dividing the participants in each group is randomized, it is less likely that they will be divided in a way where all or a majority of the boys or girls are in the same group.

There are a few ways to randomly select the participants in each group, such as using a fish bowl with a number for each participant in it or using a computer number generator. The word *random* means that each of the 20 participants had an equal chance of being selected for the experimental group or the control group. The researcher is usually aware of which group the participants are assigned to, but the participants are not aware of that information. However, there are cases when the researcher does not know the assignment of the group participants. These types of studies where neither the participants nor the researcher are aware of which is the experimental group and which is the control group are called double-blind studies.

Pre- and Posttesting

Finally, in order for us to make sure that our experiment is truly working, we need to know how sensitive our participants were before they played video games. Maybe they were not caring to begin with and whether they played violent or nonviolent video games may have very little impact. In other words, we need to pretest the participants and posttest them to figure out the actual results of our study. Otherwise we would not be able to notice any changes, would we?

This step is the final requirement for conducting an experimental design: measuring the characteristics of interest before and after we conduct the experiment. Here is an important quality of pre- and posttesting that you must remember: The pretest and the posttest cannot be different. If the tests are different, the testing itself may manipulate our outcome, so we need to give the same test to participants at the beginning of the study and at the end of the experiment. Going back to the video game example, we need to have

RESEARCH IN ACTION 4.1

ILLUSTRATION OF AN EXPERIMENTAL DESIGN

The following study is an illustration of how researchers depict an experimental design and the steps they follow to ensure all the requirements that can lead to causality in research. They depict the presence of the control and experimental groups, how the subjects were randomly assigned, the presence of the pre- and posttests, and the lapse of time. This study is trying to experiment on the effects of a chess course on improving mathematical abilities in young children.

Mathematical Problem-Solving Abilities and Chess: An Experimental Study on Young Pupils by Sala, Gorini, and Pravettoni (2015)

Source: Sala, Gorini, & Pravettoni (2015). Mathematical Problem-Solving Abilities and Chess: An Experimental Study on Young Pupils. *SAGE Open*, 5(3), 1–9. http://journals.sagepub.com/doi/abs/10.1177/2158244015596050.

The study was conducted on a total of 31 classes (third, fourth, and fifth grades) from 8 different schools of Northern Italy. The classes were randomly assigned to two groups, including 17 classes in the experimental group and 14 in the control group. The experimental group included . . . 169 males and 140 females. . . . The control group included . . . 116 males and 135 females.

Immediately we can see how the first requirement (the presence of control and experimental groups) is fulfilled. We can also see the mention of the term "randomly assigned" that completes the second requirement. The researchers also explain further how the groups were similar in terms of grade level (omitted here) and gender.

Students in the experimental group received a mandatory chess course. . . . On the contrary, students in the control group performed only the normal school activities without any chess-related activity.

The description of the experiment continues giving us details about the experiment (chess courses).

All students (both in the experimental and in the control groups) were tested before and after the intervention using the seven Organisation for Economic Co-Operation and Development–Programme for International Student Assessment (OECD-PISA) items (Organisation for Economic Co-Operation and Development, 2009), a validated instrument to assess mathematical problem-solving abilities with several degrees of difficulty (see Table 1), and a 12-item questionnaire to assess chess abilities (Trinchero, 2013; see Table 2). Time between the pre- and posttest evaluation was 3 months.

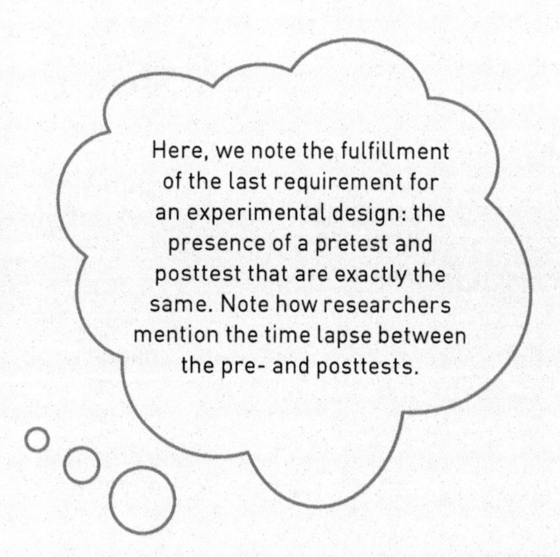

a clear set of questions on how we are measuring people's sensitivity, because if we ask people different questions at the beginning of the experiment and different questions at the end, the nature of questions alone may trigger different answers, which has nothing to do with violent or nonviolent video games.

Solomon Four-Group Experimental Design

The Solomon four-group experimental design is a type of experimental design that helps to minimize measurement errors (measurement errors are discussed at length in Chapter 5), which is shown in Figure 4.2. Sometimes researchers are concerned that when study participants take a pretest, they gain insight into the variables that researchers are measuring and change their behaviors. There are some participants who try to determine what the study is about and want to give the researcher what they think he or she is looking for. This is common human behavior, but it can bias the results of the study, which is the last thing a researcher wants to do.

The Solomon four-group experimental design protects the study from the biases of pretest because it uses two experimental groups and two control groups. From these groups, only one experimental and one control group take the pretest. The other two groups are subjected only to the intervention or the placebo and do not take the pretest. The differences in results can clearly show whether the pretest is influencing the results of the study.

FIGURE 4.2 ■ Solomon Four-Group Design

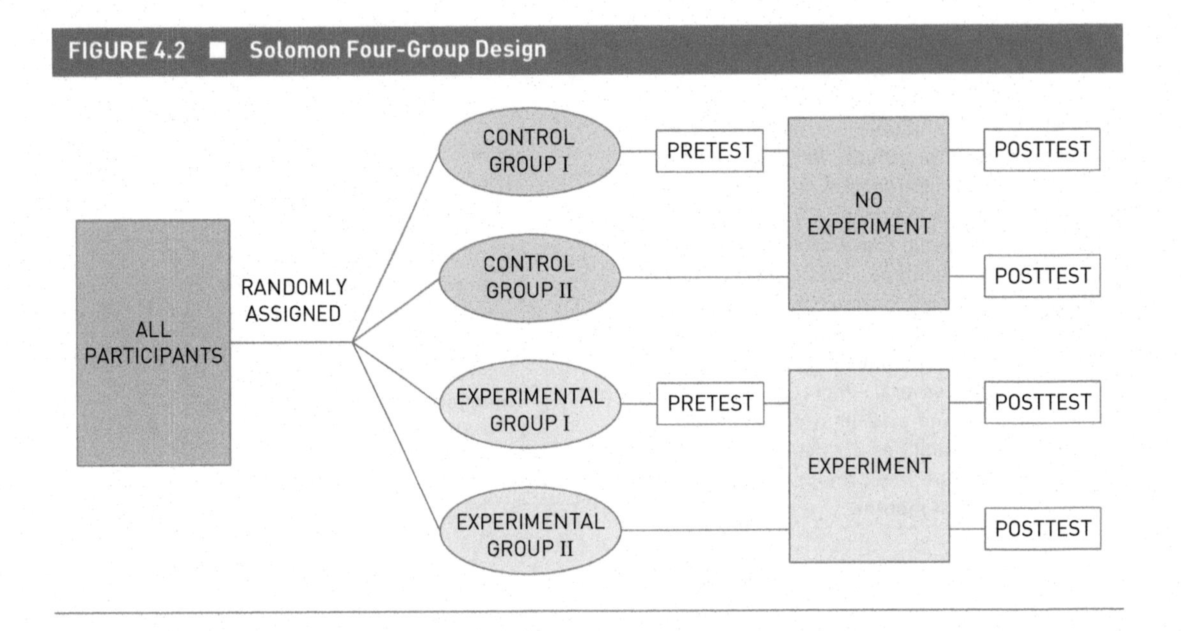

Quasi-Experimental Designs

Experimental designs have many benefits, but they are often difficult, costly, and time-consuming. Researchers have created various other ways to conduct research that are similar to experimental design, but do not fulfill all the requirements of an experimental design study. Some of these are called quasi-experimental and others are non-experimental, or pre-experimental.

Quasi-experimental designs conduct some form of intervention, testing, modification, or manipulation and examine the results. The presence of an experiment makes them quasi-experimental. Nevertheless, these types of studies are more relaxed in terms of the other experimental design requirements, such as randomly selecting participants for the control and the experimental groups or conducting a pretest and a posttest. Let us take a look at some common quasi-experimental designs.

Randomized One-Group Posttest-Only Design

Randomized one group posttest-only design has only one randomly selected group in the study. Participants are not given a pretest, but they are given a posttest to measure the results from the experiment. For example, a researcher may be interested in conducting a training writing program for entry-level college students during their freshmen year.

FIGURE 4.3 ■ Randomized One-Group Posttest-Only Design

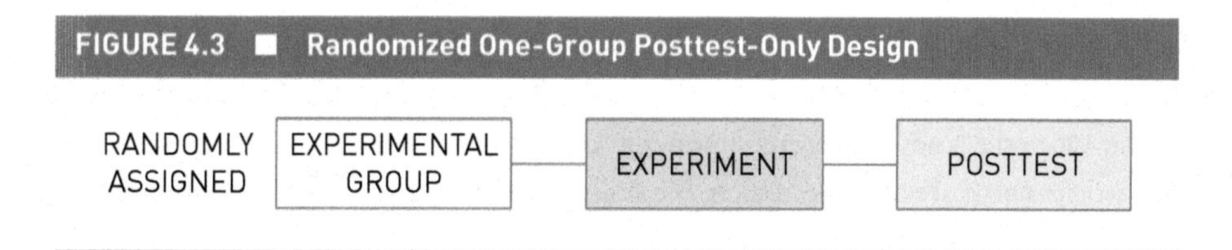

Circumstances such as cost and time may not allow the researcher to consider a control group and a testing group, but the researcher is able to randomly select a group of students from the volunteering freshmen class at a local university. These students meet weekly to get their training in writing skills, but there was no pretest done to know their starting levels of writing skills. The researcher can only collect posttest data after the intervention. So collecting posttest data and having students randomly selected in the intervention group are the only characteristics that make this study follow the randomized one-group posttest-only design.

Randomized Posttest-Only Control Group Design

A randomized posttest-only control group design includes two groups: a control group and an experimental group. Participants are randomly placed into each group. They are not given a pretest, but both groups are given a posttest. This type of design is almost identical to the experimental design with the exception of the pretest. Sometimes it is not possible to collect pretest data for various reasons. For example, a researcher may be interested in examining the impact of introducing independent life skills to widowed men who are 65 years or older. The researcher may be able to randomly select participants for

FIGURE 4.4 ■ Randomized Posttest-Only Control Group Design

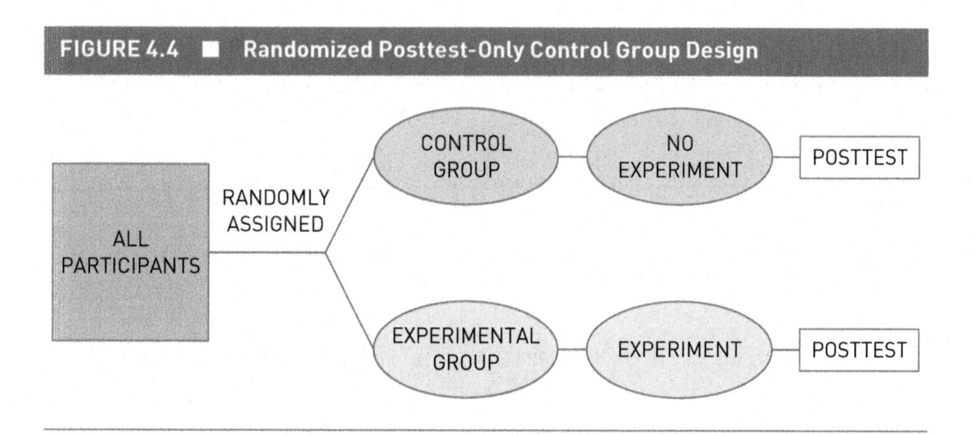

the control and testing groups, but may not be able to conduct any pretesting. During the study, one group of participant is introduced to independent life skills whereas the other is not. Once the study is over, the researcher collects posttest information on the ability of participants to use the independent life skills introduced. Since no pretest information is collected, this is a randomized posttest-only control group design.

Non-Random Posttest-Only Control Group Design

The non-random posttest-only control group design is similar to the randomized posttest-only control group design, but lacks randomization. The participants in this type of experiment are not randomly selected in either the control or the experimental group. In this case randomization as well as pretesting are missing. For example, a researcher may be interested to see the academic performance of children who follow a Montessori methodology of learning versus the traditional schooling of children during elementary school years. The researcher may not have access to pretest information of children and is not able to randomly select them in one group versus another. However, the researcher is able to collect posttest data of academic performance for children who start their 5th grade after they have been in a Montessori setting until 4th grade and compare them to students who have been in traditional school setting from the beginning. This is an example of non-randomized posttest-only control group design.

Non-Random Pretest Posttest Control Group Design

The non-random pretest posttest control group design is identical to the classical experimental design because of the presence of both the control and experimental groups

FIGURE 4.5 ■ Non-Random Posttest-Only Control Group Design

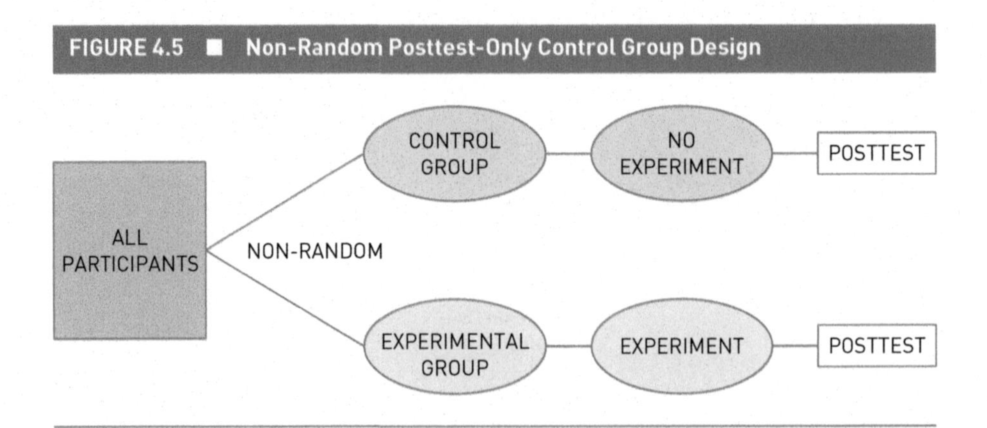

FIGURE 4.6 ■ Non-Random Pretest Posttest Control Group Design

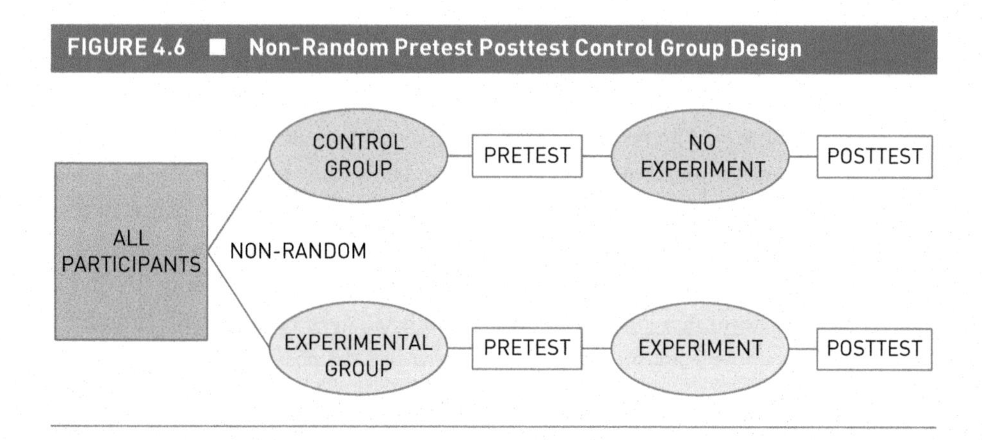

and the pretest and posttest. However, participants are not randomly chosen for each group. Therefore there is no randomization in this design. For example, a researcher is interested in the impact of the use of video during lecturing in teaching chemistry to 8th graders. The classes of 8th graders may be preselected so the researcher may not be able to randomly assign students in one group or another, but is given two different classes of students to run the experiment. The researcher introduces one group of 8th graders to the video lectures in chemistry whereas only standard lectures in the other group. In addition, the researcher collects pretest and posttest data for both the control and the experimental groups. As you can see, this is exactly like the experimental design without the possibility to randomly select participants for each group. Therefore, the researcher is conducting a quasi-experimental design, specifically non-random pretest posttest control group design.

Non-Random One-Group Pretest Posttest Design

Although this one-group pretest posttest design includes both the pre- and posttests, the study has only one group, and the participants are not randomly chosen for the study. This is another example when the researcher cannot control randomization or have

FIGURE 4.7 ■ Non-Random One-Group Pretest Posttest Design

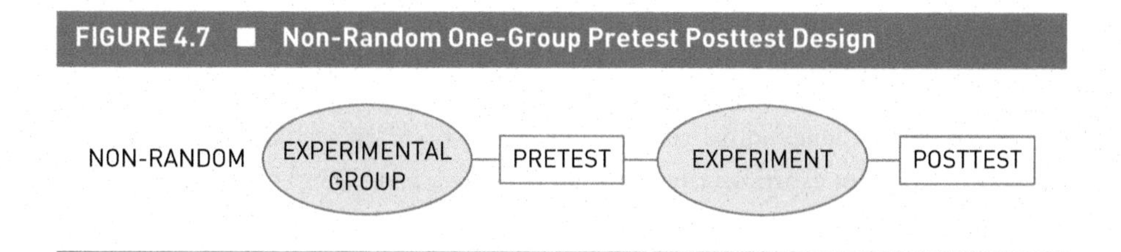

access to it, and does not have the ability to compare the results with a control group because there is only one group available. For example, a social work researcher has designed an intervention program to help victims of domestic violence cope with their traumatic experiences. The researcher has access to try out the intervention program in one of the domestic violence shelters in the area, but there is no possibility of having two groups of participants or randomly selecting the shelter or the participants. In this case, the researcher does what is possible and conducts the training program and collects pre- and post-data on various important measures, such as life skills, ability to find a job, ability to perform in a job, self-esteem, and others. Since randomization and having a control group were not possible, the researcher had the possibility of conducting a quasi-experimental, non-random one-group pretest posttest design.

These are some of the quasi-experimental designs available. What truly determines the type of design to follow is the type of study the researcher is hoping to conduct. The design accommodates the study, not the other way around. Researchers modify and create different designs to suit their studies every day. You can clearly see here why quasi-experimental studies are quite desirable because they can adjust to the study much easier than the classic experimental design.

Non-Experimental or Pre-Experimental Designs

Non-experimental designs do not include any manipulation, such as training, intervention, or drug taking. In other words, they lack the essence of being an *experiment.* These designs, however, are crucial for researchers who are exploring something innovative and will then move into conducting experiments later. The term *pre-experimental* implies that these types of studies are conducted before the researcher establishes the need for an experiment. Non-experimental designs are commonly used in all fields of science. The following are two common non-experimental designs.

Non-Random Cross-Sectional Survey Design

As you can tell from the name of this design, a non-random cross-sectional survey design refers to a survey. Participants are not randomly chosen for this type of study.

Longitudinal Cohort Study

Figure 4.9 shows a longitudinal cohort study that is based on accumulating data from participants, but not experimenting on them in any way. Any longitudinal design, such as trend or panel studies, can be non-experimental.

FIGURE 4.8 ■ Non-Random Cross-Sectional Survey Design

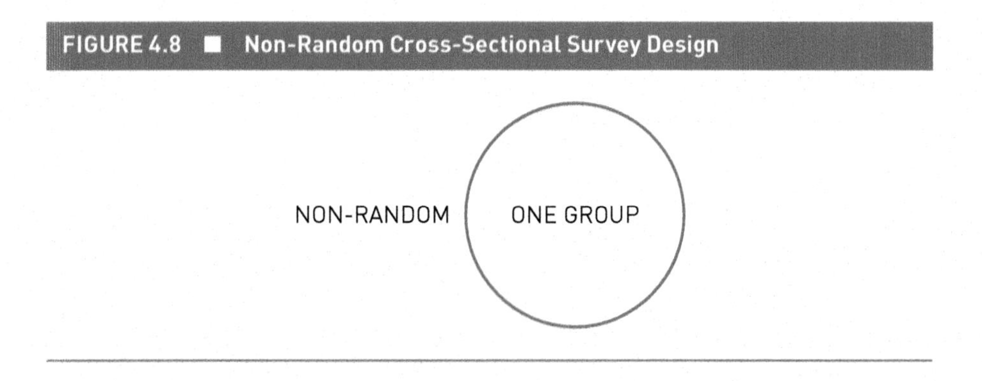

FIGURE 4.9 ■ Longitudinal Design

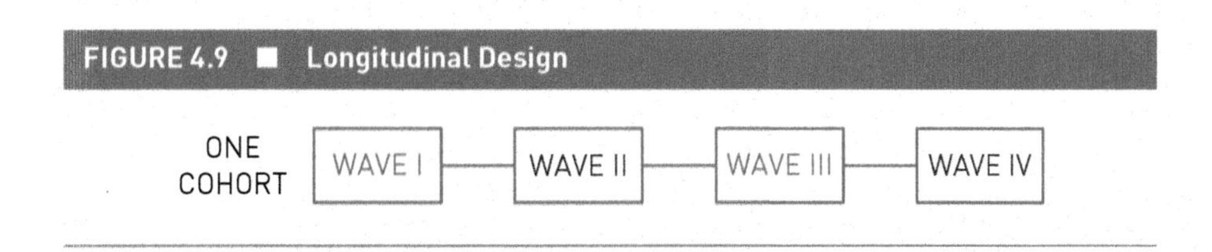

These are two common non-experimental or pre-experimental design studies. Although they do not include the specific features of experimental or quasi-experimental designs, these types of studies are as important. It is important to remember that the design serves the type of research we are conducting and not the other way around. Often, we find ourselves immersed into a new idea that was not researched before. An innovative idea could turn out to be brilliant, but we cannot take the risk to conduct costly experiments based on a new idea. For these types of research studies, non-experimental or pre-experimental designs are the best options.

Summary

This chapter introduced the idea that research design is a logical process and a researcher needs to consider every detail that goes into it. The first classification of design is between (a) exploratory, (b) descriptive, and (c) explanatory designs. Exploratory design is the type of design where (a) the relationship between variables is a new investigation not carried out by previous studies, (b) the relationship between variables is established, but the population of interest is new, or (c) pieces of the relationship between variables are founded in the literature, but the relationship link you are drawing is an innovative mixture. Descriptive designs allow researchers to focus on describing a phenomenon or understanding the details about peoples' experiences of a particular event. Explanatory design refers to studies that focus on explaining the reasons behind a phenomenon, relationship, or event.

The second classification of design refers to the frequency of data collection and attempts to determine how often the researcher will collect information from participants. Studies can be either cross-sectional or longitudinal. The studies using cross-sectional design collect the data from participants at only one point in time. Longitudinal studies collect data at different points in time. Some types of longitudinal studies are (a) panel studies, which follow and accumulate data from the same participants over a period of time; (b) trend studies, which collect data at different points in time from different participants of the same population; and (c) cohort studies, which follow a specific cohort of people over time.

A third characteristic of design is the division of studies into nomothetic and idiographic. Nomothetic research focuses on testing causal relationships between the independent and the dependent variable that can be generalized to the entire population. To establish nomothetic causality, a research study needs to establish three criteria: (1) correlation, (2) time order, and (3) non-spuriousness. Correlation assumes a relationship between variables, but does not prove causality on its own. To prove causality, the researcher needs to establish the time order and non-spuriousness. Time order determines the actual order of changes from one variable to another. The changes noticed at the independent variable must occur before changes in the dependent variable in order to establish the time order. Finally, a spurious relationship is a relationship that is explained away from a third variable that the research may not have considered. Idiographic research focuses on exploring a causal relationship for one individual or one event at a time. The purpose of this type of research is to understand the possible causes for changes on the dependent variable.

Another distinctive characteristic of design divides studies into experimental, quasi-experimental, and non-experimental. An experimental design is a study that incorporates an experiment into it and examines the results. To conduct an experimental design, the researcher needs to (a) include a control group and an experimental group, (b) randomly assign participants to each group, and (c) conduct a pretest and posttest.

A control group is a group of participants who do everything in the same way as the experimental group, but do not experience any manipulation. This group controls for the change in results of the experimental group. Random assignment means that every participant in the study has an equal chance of being selected for the experimental or the control group. Pretesting is a form of measure taken at the beginning of the experiment before the intervention, testing, or the manipulation of interest has happened. Posttesting uses exactly the same form of measure as pretesting, but is taken at the end of the experiment to distinguish any differences due to the experiment.

Quasi-experimental design studies also involve some form of experiment, intervention, or manipulation, but are more relaxed in the other requirements of the classic experimental design, such as the presence of a control group, randomization, or the pre- and posttest. Non-experimental designs are those studies that do not include any type of experiment, but are simply exploring whether an experimental study will be needed in the future on a certain topic.

Key Terms

Cohort study: a longitudinal study that follows a cohort (group sharing the same characteristic) over a period of time.

Control group: a group of participants who do everything the experimental group members do, but are not given any test, drug, intervention, or manipulation.

Correlation: occurs when changes in the independent variable cause changes in the dependent variable.

Cross-sectional study: a study that collects data only once.

Descriptive study: a study that allows researchers to focus on describing a phenomenon or understanding the details of people's experiences of a particular event.

Experimental design: a study that includes an experiment and analyzes the results.

Experimental group: the group of participants who undergo a form of experimentation, such as training, taking a test or drug, or another type of intervention.

Explanatory study: a study that explains the reasons behind a phenomenon, relationship, or event.

Exploratory study: a study used to investigate a new topic of research, a new thread of previously established relationships, a new methodology, or a new instrument of data collection, or to gain deeper understanding of a specific population.

Idiographic research: research that attempts to explain the realm of influences and details that influenced a certain phenomenon or event for an individual.

Longitudinal study: a study that collects data at different points in time.

Nomothetic research: research that seeks to prove causal relationships that can be applied to the population at large.

Non-experimental design: a study that does not include any manipulation, such as training, intervention, or drug taking.

Non-spurious: a relationship between variables that is clean of any interference of another variable, a relationship that is not explained by the presence of a third variable.

Panel study: a longitudinal study that follows and accumulates data from the same participants over a period of time.

Quasi-experimental design: a study that conducts some form of intervention, testing, modification, or manipulation and then examines the results. It is more relaxed than a classic experimental design when it comes to random assignment and pre- and posttesting.

Spurious: a relationship between variables that seems real, but is in fact explained by the presence of another variable.

Time order: determines the order of changes from one variable to another.

Trend study: a longitudinal study that collects data at different points in time from different participants of the same population.

Taking a Step Further

1. Illustrate with examples how trend studies are different from panel and cohort studies.
2. What are the conditions of having an experimental study?
3. How do we identify spurious relationships?
4. When can we state that we have a correlation between variables? Illustrate with an example.
5. What makes exploratory studies so attractive for researchers?
6. What are the advantages of conducting a cross-sectional study?

5

MEASUREMENT ERRORS, RELIABILITY, VALIDITY

CHAPTER OUTLINE

WHAT WILL YOU LEARN TO DO?

1. Recognize measurement errors and describe how to categorize them
2. Compare and contrast the inter-rater reliability, test-retest reliability, and internal consistency reliability
3. Analyze the different types of validity: face validity, content validity, construct validity, criterion validity, concurrent validity, and predictive validity

MEASUREMENT ERRORS

This chapter introduces you to the idea of measurement errors. Your study should have as few errors as possible, so it is imperative that you know how to prevent some of the common errors that can occur when designing measurement instruments. We will also discuss factors that impact the credibility of studies: reliability, validity, and their subtypes.

The existence of measurement errors is very common, so the goal is to address and minimize them as much as possible. However, we must also realize that measurement errors are likely to be present in any design. After the data are collected, most researchers understand details about their study that could have yielded better outcomes. However, being cautious and trying to think through all aspects of the study and errors that could occur from the beginning can substantially minimize these errors.

Defining Measurement Error

What is a measurement error? As researchers, we are out in the field, trying to collect data and measure concepts of interest to us. However, for various reasons, the data we collect do not represent reality because of the way we are measuring them. This is called **measurement error**. It occurs because either we are asking questions that do not truly capture what we want to capture or there is something else about the way we are collecting data that create errors.

Measurement errors can be significant or insignificant. When errors are significant, our data may be completely flawed and our work is worthless; at other times, the error may be insignificant and does not influence our results or our study. By calculating the magnitude of error in a study, a researcher's confidence in the data collected and the results can be quite strong.

Types of Measurement Errors

Measurement error can take many shapes and forms, but one way of categorizing it is as **random error** or **systematic error**.

Random Error

When the measurement error in our collected data is random and small, we have a better chance of still having high-quality data available. Random error is inconsistent and has no pattern, so our measurement may cause participants to understand and/or answer the question differently. When a small number of participants do not understand a question, accurate data are still available for analysis. However, sometimes random error is so substantial that most participants misunderstand the question and their answers do not follow an expected pattern. In this case, the random error has thoroughly damaged the quality of data collected.

For example, say we are trying to measure the frequency of students' anxiety before an exam, and we ask them a few questions as our measurement:

A. How often do you feel anxious? (1. Never 2. Rarely 3. Sometimes 4. Often)

B. How often do you experience increased heart palpitations? (1. Never 2. Rarely 3. Sometimes 4. Often)

C. How often do you experience a sense of not having enough air available when you think about something that worries you? (1. Never 2. Rarely 3. Sometimes 4. Often)

These three questions were intended to measure test anxiety. However, research shows that people understand the response terms differently. In fact, a study that examined how people interpret qualifiers like *often* found that people perceive these terms differently (Barnes, Cerrito, & Levi, 2003). Barnes et al., (2003) conducted a large study in an urban university to understand how people interpreted qualifiers like *often*. They found out that for some people the word *often* may describe something that happens daily, whereas for other people it may mean something that happens weekly, biweekly, monthly, or even annually (Barnes et al., 2003). The same subjectivity of terms applies to words like *rarely*, and *sometimes*. Therefore, we may end up with answers that do not follow a pattern and have problematic data because of severe random error. It helps to be more specific with either the answer choices or the question when they include vague qualifiers like *often*.

In our example about testing exam anxiety, we could leave the questions worded exactly as they are, but offer answer choices that are more specific: (a) daily, (b) twice a week, (c) weekly, (d) biweekly, (e) monthly, (f) almost never. This way, we would minimize the random error that may result from misinterpreting the qualifier *often* in the question. To minimize the random error even further, we could remove the word *often* entirely from the question and ask about the frequency of feeling anxious, having heart palpitations, or lack of air.

Fortunately, random error is usually minimal and, in most studies, applies to a handful of answers where participants may not have been paying attention when answering or they interpret the question differently from what was intended. Our exam anxiety example is a problematic random error of the measurement, but we are usually aware of these types of errors and take them into account before data collection. To distinguish random error from other types of measurement error, we should remember that random error has no consistent pattern, and it implies that our measurement may be open to interpretations from various people. When we design our data collection tool, we try to minimize random error by considering our questions wisely.

Random error can also occur in cases where our questions may be clear and specific about what we are measuring—meaning that there is only one way of interpreting it—but the answers may be highly socially desirable. Social desirability is a common problem in research because participants may feel like their answers must represent what they *should* be doing rather than what they *actually* do. In those cases, some people may be inclined to avoid answering truthfully, and we may have a slight problem with random error. Socially desirable questions may be personal questions. For example, if parents are asked about how many hours their children watch television, play video games, or spend time with electronic gadgets, they may feel uncomfortable to answer truthfully because some parents may perceive the answer as a reflection of their own parenting qualities. Concurrently, people may be skeptical in telling the truth about how they feel about their neighbors, whether they read self-help books, whether they drink too much, or whether they use illegal drugs and other similar questions.

Researchers have come up with some ideas on how to reconstruct questions in such a way that people may be less likely to think of the answers as negatively reflecting on them and may be more willing to answer truthfully. For example, let's return to the case of parents being asked about their children's electronic usage. If parents are first asked about how difficult it is to stop their children from using electronics and are then asked approximately how many hours their children spend online or playing video games, they may be less likely to feel like their children's use of electronics is a reflection on parenting quality, but rather a societal problem.

Systematic Error

Systematic error, on the other hand, is far more problematic because it indicates that our measure is not accurately measuring our concept, but is systematically perceived as something else from participants. Systematic error can be related to the measurement we are using, but it can also relate to the way we are collecting the data or other environmental factors unrelated to the measurement. A systematic error related to measurement would be when our instrument of data collection is flawed so that the results are either too high or too low. An error is also systematic when answers from the majority of participants display the same type of error.

For example, say we were interested in measuring people's religiosity and we created two questions:

A. Are you religious? (1. Yes 2. No)

B. How often do you go to church? (1. Daily 2. Weekly 3. Monthly 4. On holidays 5. Never)

Can we measure religiosity by asking people how often they go to church? What are we implying with that question?

You may have already noticed what is wrong with measuring the construct of religiosity with these questions, but for the sake of argument, let's continue. The data were collected in a neighborhood in Queens, in New York City, and our participants were mostly Muslim and Jewish people who lived in the area. The results showed something strange: Participants claimed that they were religious, but they rarely set foot in church. Does this mean that they were somehow alienated by their local church? Were they very busy and could not afford the time to go to church? No. They were going to other religious institutions, such as synagogues and mosques, and it was our question that created a systematic error in our data. Therefore, a systematic error is an error that can lead to a specific, different answer from the focus of the research. Systematic error can be caused by the measurement, but sometimes it is present in the environmental or social circumstances.

Let us examine a systematic error that is present in the environment rather than in the measurement. In most U.S. colleges, students evaluate the teaching quality of their professors at the end of the term. Students can voice their opinions about how well prepared the professor was for class, his or her strengths or weaknesses, and so on. However, if students complete these evaluations right after they receive a bad grade on their paper, they are likely to commit systematic error by poorly evaluating the professor just because they feel that the grade they received was unfair. As you can imagine, the grade one receives in the class or the perceptions of fairness or unfairness about the professor is unrelated to his or her quality of teaching.

RESEARCH WORKSHOP 5.1

HOW TO MINIMIZE MEASUREMENT ERROR

In practice there are a number of known ways to minimize measurement error from interfering with the quality of our dataset. We are especially concerned with minimizing systematic error, but trying to minimize both kinds of errors is always a good idea. Here are some steps that may be helpful to your study:

1. Conduct a pilot study—or a trial data collection. Just as if you were rehearsing before an important presentation to a large audience, you may consider rehearsing your data collection process by getting together a small sample of participants. You may ask friends and family members in this case because you are not going to use the information from this trial—you are simply attempting to minimize measurement error. Once the trial of data collection is conducted, you can ask your participants for feedback. They will likely tell you whether a question was confusing to them or whether they believe an additional question should have been added for better results.
2. If you are collecting information by hand or paper surveys/questionnaires, and then later entering the answers in a spreadsheet format, you may consider asking a friend to reenter the data in a separate spreadsheet. Later you can compare the two spreadsheets and see if there are any inconsistencies. This can help minimize random error, or the error caused by mistakes in data entry, and although random error is not very problematic, it is a good idea to do the best you can to ensure higher quality of data.
3. Another quite useful technique is adding questions that measure similar constructs—this is especially important for constructs that are crucial to the study. By having a number of questions that are measuring the same construct, you are likely to avoid systematic error that can be caused from a common misunderstanding of one single question. This way, you can at least be certain that your dataset is clear of confusing questions.

RELIABILITY

As researchers, we want our study to be as perfect as possible. Naturally, we have worked very hard in choosing all the characteristics of design, have read and synthesized the literature review, have broken down our constructs into variables, and are now trying to prevent errors during data collection. Making sure that our data collection tool does not have systematic errors or grave random errors is the first basic step. Next, we need to ensure reliability and validity of these measurements. Both terms relate to the strength of our study in terms of errors.

Reliability is the consistency in our measurement. We can state that our data collection instrument is reliable if it will yield the same results even if used with different subjects, different populations, and/or different settings. This means that the measurement tool will be perceived the same way by all types of participants. Reliability refers to the level of clarity in the tool.

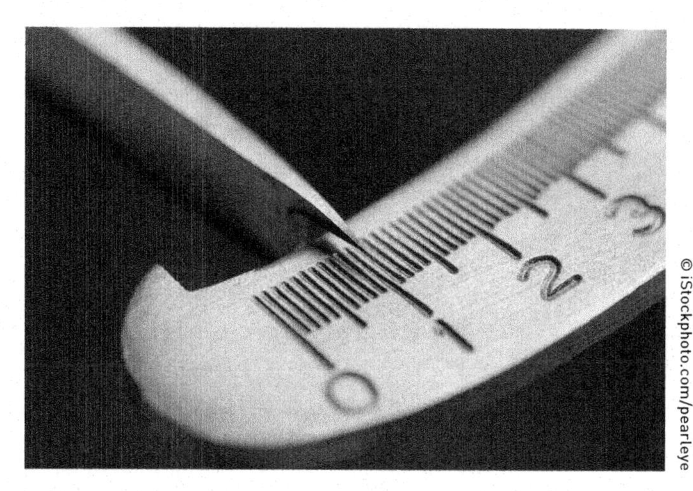

Are we truly measuring what we want to measure?

A good comparison to a reliable instrument is a painting from the era of realism. A realistic painting of a horse is showing a horse using as many details as possible. There is no mistaking it for a dog or elephant or any another animal. Anyone looking at this realistic painting will recognize that it is a horse. In other words, the painting is reliable in portraying a horse. An abstract painting, on the other hand, is not reliable because people see it differently and interpret it different ways. They may not agree with each other in terms of what they perceive in the painting. For now, we want our instrument of data collection to be as close to a realistic painting as possible to increase the reliability of our study.

When considering reliability, it is important to remember: *Reliability does not mean accuracy.* The accuracy of data does not pertain to reliability. Reliability is simply making sure that the instrument we are using to collect data is truly measuring what we want it to measure. **Accuracy** in data collection refers to participants' answers about a topic and whether or not the answers are free from error. Some concepts are more difficult to measure than others because people perceive them differently.

One such concept is the measurement of time. If we ask participants approximately how much time they spend performing a specific activity by clearly stating our question, our measurement may be reliable—there are no mistakes about what we are asking—but the answers may not be accurate. People do not measure time simply in terms of the minutes or hours they spend doing an activity; they also factor in how they feel about the task at hand. For example, you may be sitting in a very engaging lecture and feel like only 10 minutes have passed when an entire hour has passed. Similarly, you may experience a boring, tedious lecture and feel like you have been sitting there for hours on end when only a few minutes have gone by. Therefore, while a measurement tool may measure what it intended, the results may be inaccurate. Perhaps we would get more accurate answers if we ask participants to specify how much time they have spent in a particular task as well as specify their likability of the task. The two answers together may result in a more accurate measurement of time.

Now we know what reliability is and can distinguish it from accuracy, but how do we truly know that our instrument is reliable? We obviously may not have sufficient time to collect data from different populations, subjects, and in settings to state that the instrument is reliable. In reality, we feel lucky to have a handful of participants who agreed to volunteer in our study, so how do we handle reliability? Three popular ways of ensuring reliability are (1) inter-observer or inter-rater reliability, (2) test-retest reliability, and (3) internal consistency reliability.

Inter-Observer or Inter-Rater Reliability

In the case of **inter-observer** or **inter-rater reliability**, there is more than one researcher present in the data collection. All the observers, or raters, keep separate ratings of the data and then compare responses to determine a degree of consistency. This procedure is used in various types of studies, but it is more common in qualitative studies when the researcher may need an additional researcher to code or collect the same material.

For example, if a researcher is observing participants in a new culture or environment, it may help if another researcher is also working on the same project and observing at the same time. The two can compare their notes and see how consistent they are in what they are observing. This procedure helps increase the reliability of a study. This procedure is especially helpful in case studies. A researcher may read and code a large number of case studies on people with substance abuse problems. In such cases, it is helpful to prepare a list of codes beforehand regarding how the cases are being sorted, share the list with another researcher or rater, and, once the work is completed independently, compare the results. Good consistency between the raters indicates high reliability of the study.

Test-Retest Reliability

Test-retest reliability ensures that our data collection tool consistently yields the same results regardless of the passing of time. This means that data are collected at one point in time and are collected again at a later point in time. Often the test-retest reliability is conducted using the same group of participants when stability over time or the setting of the data collection may have influenced the reliability of the study.

It is important to continue to keep in mind that all design features, data collection, or reliability measures serve the study—not the other way around. Therefore, if a study is concerned with changes over time, test-retest reliability would be a better approach than inter-observer reliability. Test-retest reliability is especially important in instruments that measure data on different types of disorders or health problems.

For example, imagine that a researcher wants to determine whether yoga poses are helpful to people trying to fight stress. This researcher may choose a specific set of yoga poses and tries an experiment with a group of participants. Over the course of a month, the participants engage in this set of yoga poses in an indoor high-quality gym with forest and waterfall simulations in the background. The researcher measures their levels of stress before and after exercise by asking them to answer some questions and then measures their cortisol levels—stress hormones. Once the month is over, the results seem to show that yoga poses did lower participants' level of stress, but it is unclear whether going to a local gym or doing the same poses at home or the office would have the same results. What if the high-quality gym and the background simulation noises were the relaxing factors rather than the yoga poses? Reconducting the study in a regular gym a few months later will increase the reliability for this study through the test-retest method.

Internal Consistency Reliability

Often used in surveys and questionnaires, **internal consistency reliability** is a useful tool that increases the reliability of studies. Before explaining what internal consistency is, think about your experiences in completing a questionnaire or a survey. Can you think of any times when you answered a question and then, a few questions down the list, saw the same question with a slightly different structure? You probably thought to yourself, "I answered this before!" When we are in a hurry, those similar questions may annoy us. This is exactly what internal consistency reliability is. To make sure that the participants truly understood the question and their answer is not due to an error, the researchers include a few questions that measure exactly the same variable. They may or may not structure them differently, but the question will ask about the same phenomenon. Often these similar questions are put in a similar scale, so the researcher can easily identify their consistency.

For example, let's imagine that a test is attempting to capture how depressed a patient is feeling at the moment by asking the patient the following questions:

A. How depressed do you think you are on a scale from 1 to 10, where 1 is the lowest or not depressed at all and 10 is the highest or very depressed?

B. How sad do you perceive yourself to be on a scale from 1 to 10, where 1 is the lowest or not sad at all and 10 is the highest or very sad?

C. How gloomy do you think you are most days on a scale from 1 to 10, where 1 is the lowest or not gloomy at all and 10 is the highest or very gloomy?

RESEARCH IN ACTION 5.1

DETAILS ON STRENGTHENING A QUESTIONNAIRE

In this article, we can look at the details about strengthening a specific widely used questionnaire. This questionnaire is used to identify emotional and behavioral problems in children and adolescents. The article goes into the details of reliability of the questionnaire over time and in different countries.

Source: Menon (2014). The Strengths and Difficulties Questionnaire: a pilot study on the validity of the self-report version to Measure the Mental Health of Zambian Adolescents. *Journal of Health Science, 2* (1), pp. 127–134.

The SDQ is a brief screening measure that is increasingly being employed for the purpose of identifying behavioral and emotional problems in children and adolescents. The SDQ has been used in clinical and service activities and over the years has been recognized as a widely used measure of child and early to mid-adolescent mental health.

This paragraph is taken from the introduction and gives us a sense about what the SDQ is. It is simply an instrument (questionnaire with 25 questions) that identifies emotional and behavioral problems of children.

RELIABILITY OF SDQ

Cronbach alpha coefficients were calculated for the total scores . . . to determine internal consistency. . . . Internal consistencies for emotional symptoms and the pro-social scale were adequate. Internal consistency for hyperactivity was low and very low for conduct disorders and peer problems.

From the results section, we see that the SDQ has shown internal consistency of items for emotional symptoms. However, internal consistency for hyperactivity, conduct disorders, and peer problems is low, meaning that the instrument is reliable in measuring emotional symptoms, but not hyperactivity, conduct disorders, and peer problems.

Studies have explored the reliability of SDQ in various cultures. . . . Similar to the findings of this study, the Chinese study reported low internal consistency for conduct problems. . . . And a study of Norwegian teenagers . . . found that conduct disorders had the lowest internal consistency.

This last section taken from the discussion compares the results with other studies in other countries. Here we also note that the internal consistency of the instrument has been low in other studies from other countries in identifying conduct disorder problems among children and teenagers.

All these questions measure the same thing in the same scale. If all the ratings that the patient chooses are identical or similar, then we can state that our questionnaire has high internal consistency reliability. If the ratings vary greatly, with the patient saying he or she does not feel depressed at all, but portrays himself or herself as very gloomy and somewhat sad, then our internal consistency reliability may be low and we need to change the wording of the question so it works better in capturing the same concepts.

VALIDITY

Being consistent in our measuring or having high reliability does not mean that our results are valid. **Validity** refers to the ability or the potential of our data collection tool to capture and measure the construct or the phenomenon that we are interested in measuring. Are our questions/tests/other measures reflecting the real meaning of the concept under consideration? Regardless of our study's focus, we must ask ourselves how we will scientifically measure it. Once we have a clear idea about how to measure our concept, we must make sure that the way we have decided to measure a specific construct actually does so.

Sometimes this is an easy task because of previous research conducted and an agreement between researchers on how to measure the constructs. At other times, however, the concepts may be vague and definitions may be interpreted in various ways from different people. It is also crucial to differentiate here between reliability and validity,

Could acculturation be measured by our legal status of working and living in another country or is there something more to it?

as an instrument can be reliable without being valid, but not the other way around. To have validity, we take for granted the reliability of the instrument. In other words, validity cannot exist without the presence of reliability, but that is not the case for reliability alone (Kimberlin & Winterstein, 2008).

Our task is suddenly more complicated than we thought initially. A good example is trying to measure a concept like *acculturation.* Acculturation is the degree of embodying and understanding the new culture for immigrants. So how can we really measure if someone is completely immersed in the American culture? Being acculturated does not mean that person is assimilated or has forgotten his or her own culture; it just refers to the level of capturing the culture of the place where immigrants live. If we were interested in knowing how long it takes for an immigrant to become acculturated into the American culture, it would be easy to measure the length of time, but quite difficult to try to measure acculturation. Does acculturation mean that you celebrate Thanksgiving, or eat typical American food like burgers, or send thank you cards, or dress differently from your own culture? Well, one issue with these types of questions is that they cannot be generalized to all Americans, let alone immigrants. Another problem is that they do not seem to measure culture at all. There are many different measures for acculturation, which sometimes target specific populations, such as Hispanic or Afro-Caribbean communities. Most of these instruments have been tested multiple times by numerous researchers to ensure validity.

ETHICAL CONSIDERATION 5.1

IMPORTANT WHEN SELECTING THE APPROPRIATE INSTRUMENT

In selecting the appropriate instrument for a study, researchers need to consider participants' rights to have their own values, attitudes, and opinions. These rights need to be incorporated when the instrument is constructed so that participants will be comfortable sharing their opinions without feeling judged. This is good for the quality of the study, and it protects participants' values and attitudes. Sometimes researchers may make participants feel uncomfortable, causing them to fear that their opinions may be misinterpreted or misused. That is why researchers need to make sure that the instrument is not discriminatory in any way or makes participants feel inadequate regarding their morals, attitudes, or opinions.

Face Validity

Face validity means that a researcher or a group of researchers have decided to classify the measurement instrument or tool as an accurate measure of the construct in question. Therefore, rather than questioning or testing the measurement tool used to collect data, the researcher subjectively considers it an accurate form of measurement and interprets the results or follows up with some other questions. In other words, face validity refers to a subjective assessment of the validity of a measurement instrument made by researchers. It does not imply that the measurement is valid; it implies that for the sake of the researcher's discussion or viewpoint, the measurement is considered valid. It is widely used when researchers collect secondary data, or data that was previously collected for another purpose or study.

Face validity is often used because we cannot always confirm that a measurement tool is valid and we need to move on to other research. Therefore, when a researcher states that he or she is taking a specific study at face validity, it means that he or she will not question how the study was conducted or whether the measurements were accurate, but will assume that they were and will move on to analyze the results. In such cases, the researcher will state that the interpretation of results is taking the findings at face validity.

Content Validity

Often researchers attempt to capture constructs, such as intelligence or level of proficiency in a subject like mathematics or other areas. In such cases, the researcher is concerned with **content validity**. Content validity refers to the extent or level that a measurement tool captures all the aspects of the construct that is being tested or measured. This helps us to ensure that our measure truly includes all the aspects that make up intelligence, or mathematics, and so on. For example, the majority of final exams usually include all the concepts that the professor thinks will indicate students' level of knowledge gained during the term. For the measure (final exam) to have high content validity, it cannot include only questions from one lecture or two chapters of the textbook. It must have a wide range of coverage of the material taught during the term, otherwise its content validity is low.

Construct Validity

Construct validity is similar to content validity in that it measures the level or the degree that a measurement is able to truly measure the construct in question. Recall our discussion in Chapter 2 of the operationalization of constructs, or turning a construct into one or more variables. Some constructs are simple and can be easy to operationalize (for example, ethnicity can be measured by asking people to check a box according to their ethnicity). Other constructs are more difficult and require attention, time, and research to be able to operationalize with confidence. Construct validity is related to the degree that our operationalization of a construct is accurately done.

For example, if we are interested in measuring the level of a person's happiness, we may face some challenges. What is happiness? Are we referring to a self-satisfied feeling, or are we referring to feeling excited and energized? Do we simply ask people how happy they feel? Do we look into how optimistic people feel, or is it about their perception of the amount of control they have over their lives? For a complicated construct like happiness, we would need a few questions to begin to have some construct validity. The same set of questions should measure a sense of happiness among different groups of people. We could ask survey participants to rate a few statements on a scale from 1, strongly disagree, to 4, strongly agree, for the following questions:

A. I am happy with my life.

B. I think the world is a beautiful place.

C. I enjoy life as much as I can.

D. I consider myself to be lucky.

E. I generally have a positive outlook on life.

After the participants have completed their surveys, we can create a more specific scale of happiness. Since we have five questions on a scale from 1 to 4 where 4 indicates the most happy and 1 indicates the least happy, it is clear that if someone answered the minimum in all questions (strongly disagree = 1), we would have a minimum score of 5 for the least happy and a maximum score of 20 (strongly agree = 4) for the happiest. Therefore our scale would vary from 5 to 20 where the highest number reflects the happiest of participants.

For most constructs, we would revisit the literature and study the measurements that have already been created from previous studies. We may alter the questions and/or add others that may be more appropriate, but we should use the measurements that researchers have used for many years as a foundation or template for the survey. These measurements have been tested among different samples of the population and have more construct validity than a tool that we are creating from scratch. This does not mean that we are not supposed to create measurements from scratch. It simply means that it is easier to use proven measurement tools and improve on them.

Criterion Validity, Concurrent Validity, and Predictive Validity

We are concerned with **criterion validity** if we are creating a new measurement tool for a construct that already has a measurement tool in place. To have criterion validity, we need to establish both **concurrent validity** and **predictive validity**. When discussing construct validity, we discussed reading the literature to see how other researchers have measured the constructs that we are interested in. Criterion validity refers to those cases when we are not satisfied with the way the construct has been measured—the measurement tool may have some flaws or it does not apply to our specific population—so we decide to create a new measurement for a construct that has already been measured. In that case, we need to be clear that our new measurement tool will be compared to the old measurement tool, so we know it has high criterion validity.

For example, say we want to create a new measurement for mathematical skills. We may believe that the old instrument is insufficient, so we work hard to create a new test that should accurately measure the mathematics skills of an adult. To have high criterion validity, we must check how compatible the new test is with the old test. After all, we know that the old test was working fine, but perhaps not as perfect as we want it. We are claiming that the new test is an improved measurement of the same old test.

The first requirement is that the new test works as well as the old test. Therefore, comparing the new test with the established measurement tool and examining the matching results is mandatory. This type of comparison is called concurrent validity. Traditional mathematics tests have a number of questions that participants are expected to solve and these tests are considered accurate in measuring mechanical mathematical skills. One can argue, however, that another way of measuring mathematical skills would be a measurement of logical questions around numbers rather than mechanical skills. We take for granted that the logical questions would also yield high results in mechanical skills, so the new test will have concurrent validity with the traditional one, but even go a step further in terms of accuracy.

For example, in a traditional test one may be given an exercise that looks like this: $20*(8 + 2) \div 5 = [...]$. The participant calculates the results and finds the answer. In our improved test, we may formulate the question this way: "Please determine how to modify (such as by adding parentheses) the following exercise to yield a result of 60." Exercise is $20*8 + 2 \div 5 = 60$. You can see here that if someone answered correctly to the nontraditional exercise then he or she will likely answer correctly to the traditional one as well because the second exercise has an additional logical layer and requires the knowledge to understand the traditional exercise. It is also likely to correctly answer the traditional exercise, but incorrectly answer the nontraditional one. Once we administer both tests to a group of participants and see that almost everyone who correctly answered the nontraditional test also correctly answered the traditional one, we can claim a high concurrent validity between the tests. We may find that a small number of participants answered correctly the traditional one, but not the new test and that also proves high concurrent validity because the new improved test is claiming to measure logical skills in addition to mechanical mathematical skills, so this type of discrepancy is expected. On the other hand, if we find that most participants answered correctly the new nontraditional test, but incorrectly the traditional test, we are facing low concurrent validity because our new test seems to measure something else unrelated to the traditional test.

If the new test has high concurrent validity, we will test it a few times with different population samples. We will need more than one population to establish predictive validity. If we can claim concurrent validity with one sample, we need a number of groups of participants to test both exams multiple times. If our results over time still show that the new mathematics test is better than the old test, it establishes high predictive validity. Once both these requirements are fulfilled, as researchers, we can be confident that we have high criterion validity.

RESEARCH WORKSHOP 5.2

HOW TO SELECT THE PERFECT VALID AND RELIABLE INSTRUMENT FOR A STUDY

When planning your research project, you want to make sure that the instrument you are using to collect data is both reliable and valid. The following are some steps to select the best possible instrument:

1. Have a clearly defined construct of what you are measuring. Knowing what variables you will be measuring should be the very first step, before you start looking for existing instruments.
2. Conduct a literature search using academic databases and identify a list of instruments used for similar constructs. Once you have identified these studies, tease out the instruments that measure the constructs.
3. Start evaluating the constructs by making sure that they truly answer the questions relating to your constructs. Ignore the parts of the instruments that may be irrelevant to your construct and create a separate list with the parts of the instrument relevant to your constructs—even if they include similar questions.

4. Read the literature to determine if the instruments have been tested for reliability and validity. Studies that use well-established, reliable, and valid instruments will include that information in the study.
5. Finally, make sure that the instrument is acceptable to the population from whom you will collect your data.

Difference Between Reliability and Validity of the Measurement Instrument

The difference between reliability and validity is depicted in Figures 5.1 and 5.2. As you can visually see here, the pattern of flower petals in Figure 5.1 is reliable. It produces a petal every single time with different populations—illustrated in different sizes and colors. So we can rely on the pattern to produce the same thing over and over, but the measure is not accurate—or has no validity—because it is not truly creating a flower. In other words, it is producing a petal, but not creating a valid flower. Petals on their own right are not flowers. In Figure 5.2, we can clearly see how both reliability and validity meet together and can create the flower we need. The reliability allows for petals to be created every time and accuracy or validity puts them together to create the actual flower.

FIGURE 5.1 ■ Reliable but Not Valid

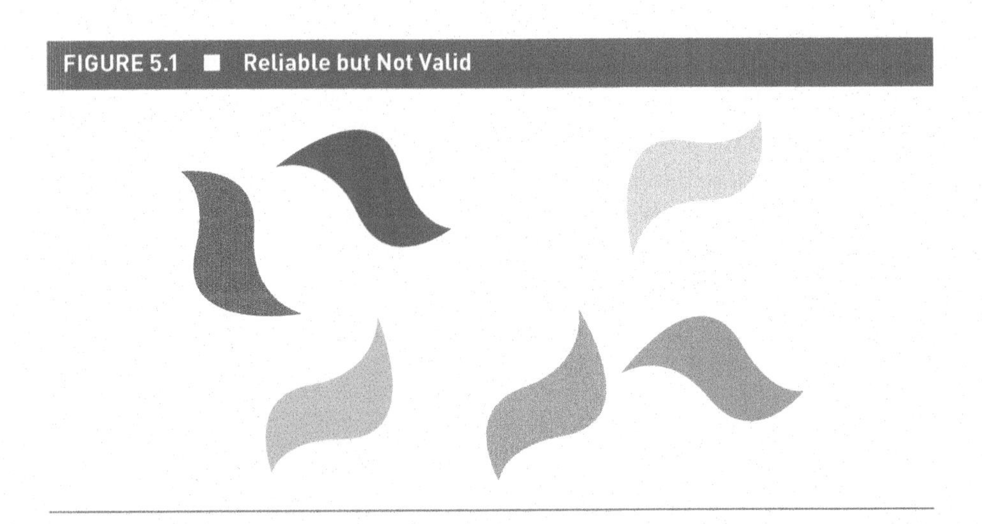

FIGURE 5.2 ■ Reliable and Valid

Summary

A measurement error is the discrepancy between the reality of the data and the data collected because of the way the construct was measured. Measurement errors can happen because of the way a question is asked by the researcher or understood by the participants. These errors can be significant or insignificant.

Measurement errors can be random or systematic. Random measurement errors refer to problems in the measurement that are misinterpreted by some, but not all the participants. Systematic measurement errors, on the other hand, show a clear pattern in the measurement that was misunderstood, misinterpreted, or mistaken by the majority of the participants. These errors are problematic because the data collected can have inaccurate information that cannot be used for data analysis.

To have the perfect measurement instrument for our study, we try to have a reliable and valid measurement. Reliability refers to the consistency of measurement. We have a reliable instrument when the same question is asked to different populations repeatedly and it provides the same consistent results. Validity means that the instrument is truly measuring what we want it to measure and ensures the instrument is an appropriate measure of our construct. We cannot have validity without having reliability first.

Three ways of making sure our data are reliable are inter-observer or inter-rater reliability, test-retest reliability, and internal consistency reliability. Inter-observer reliability asks for the presence of more than one researcher in the study. Two pairs of eyes are better than one, so when we have two or more researchers examining an instrument and handling the data separately, we can ensure inter-observer variability. This type of reliability is most common in qualitative studies when the data collected can be an outcome of different interpretations.

Test-retest reliability is often used for many measurement instruments and collects data from different populations over time. Every time we use the same instrument, we can learn more about its reliability. This is especially important in instruments that measure data on different types of disorders or health problems.

Finally, internal consistency reliability means asking the same question a few times within the same instrument. We are trying to make sure that the participant truly understood the construct we are trying to capture. By putting the construct into different questions within the same instrument, we allow the participant to provide answers to the same construct a few times. If the answers remain consistent, we can confirm that the instrument has high internal consistency reliability.

Just as there are different types of reliability, there are also different types of validity. Some common types are face validity, content validity, construct validity, criterion validity, concurrent validity, and predictive validity. Face validity means that the researchers decide to consider the data as accurate and not look closely into the accuracy of the instrument. They want to interpret data as they are, at face validity.

Content validity is concerned with the content we are trying to measure and is widely used in education. If we are trying to test a participant's ability to do mathematics, write well, or understand biology, we need to be certain that our questions are an accurate measure of these skills. Construct validity is similar to content validity, but refers to accurately measuring the construct we want to measure. There are many constructs that are important to researchers, and only by conducting a thorough literature search will we be able to ensure construct validity and accurately measure the construct.

Criterion validity refers to creating a new instrument based on the old instrument that measures a construct. We may believe that the old instrument is lacking or missing something, so we would like to improve on the instrument by using some of the old measures and adding new ones. For criterion validity to occur, we need to have both predictive validity and concurrent validity. Predictive validity is the ability of the new measurement to achieve results that are as good as or better than the old measurement. Concurrent validity matches the new instrument with the old instrument on the same constructs and ensures that the new instrument includes whatever the old instrument was measuring accurately. If we have both concurrent and predictive validity, we can see how compatible the new instrument is and ensure the presence of criterion validity.

Key Terms

Accuracy: refers to whether or not participants' answers are free from error.

Concurrent validity: validity that occurs when comparing a new test with an established measurement tool and examining the matching results.

Construct validity: validity that is related to the degree that a researcher's operationalization of a construct is accurately done.

Content validity: validity that refers to the extent that a measurement tool captures all the aspects of the construct that is being measured.

Criterion validity: validity that occurs when creating a new measurement tool for a construct that already has a measurement tool.

Face validity: validity that occurs when a researcher or group of researchers decide to classify the measurement instrument or tool as an accurate measure of the construct.

Inter-observer or **inter-rater reliability**: a process of determining reliability in which there is more than one researcher present during data collection. The observers or raters keep separate ratings of the data and then compare responses to determine a degree of consistency.

Internal consistency reliability: a process of determining reliability in which the measurement tool includes multiple questions on the same construct; if participants answer the questions similarly throughout the tool, there is high reliability.

Measurement error: occurs when the data collected do not represent reality because of the way they have been measured.

Predictive validity: validity that occurs when a measurement test produces the same results over time.

Random error: an error in measurement in which a small number of participants do not understand a question, but accurate data are still available for analysis.

Reliability: consistency in measurement (i.e., it yields the same results even if used in different subjects, different populations, and/or different settings).

Systematic error: an error in measurement in which the tool does not accurately measure the concept and is perceived incorrectly by most or all of the participants.

Test-retest reliability: a process of determining reliability in which the data collection tool consistently yields the same results regardless of the passing of time.

Validity: the ability or potential of the data collection tool to capture and measure the construct that the researcher is trying to measure.

Taking a Step Further

1. What is the relationship (if any) between validity and reliability?
2. How do we increase reliability without risking validity?
3. What measurements do we take to ensure high validity?
4. Which type of error is most problematic for our data and what can we do to minimize it?
5. How is test-retest reliability different from predictive validity?
6. What are some circumstances that make us take a research study at face validity?

SAMPLING

CHAPTER OUTLINE

WHAT WILL YOU LEARN TO DO?

1. Explain the purpose of sampling
2. Compare and contrast probability and non-probability sampling
3. Describe the types of non-probability sampling
4. Summarize the types of probability sampling

DEFINING SAMPLING

You are becoming more familiar with the research process and by now, you must be feeling confident about your literature review and the decisions you made about your design. It is time to consider the population from which you will collect the data.

A sample is a small part of an entire population. Ideally it is representative of all the characteristics of the population.

Some researchers look forward to this moment, and others find it slightly disconcerting because so many questions arise. How do we approach the population selection scientifically? Is it possible to have an entire population participate in a study? Is a subset sufficient to provide the data needed? Can we ask our friends to participate? If we have only a representative group of people participating in our study, how can we draw conclusions about the entire population?

All these questions are related to **sampling**. Sampling is the procedure used by researchers to select a subset of the population—called a **sample**—that can be used to conduct the scientific study. By following sampling procedures, researchers increase the likelihood of getting a representative subset of the population. The **population** is the entire group of people that are the focus of the study. If a study is focused on investigating homeless people, the entire population of homeless people is the population for this study. If a study is focused on investigating shoppers at Trader Joe's, the people who shop there make up the complete population for this study. Therefore, a population can be as small as the number of patients in one hospital or as large as the entire world population, depending on the study's focus.

Researchers approach the population in different ways conditioned on what fits best with their study. It is impossible and even counterproductive to include an entire population in a study. This is where a sample will suffice, especially if this group is representative of a larger population. Now, if we were to sample a cookie from a box we bought at the grocery store, we know it will be an exact representative of the entire population—there is no need for us to eat the whole box. But humans come in all varieties so we cannot make the same claim when it comes to people. To do the best possible research and have a sample of people that is representative of the entire population, we need to follow the rules of sampling.

Although there are many types of sampling, most of them begin by developing a sampling frame. The **sampling frame** is a list of the entire population of a study. So if we were to conduct a study on the effects of social capital among the homeless population in Texas, our sampling frame would be a list of all the homeless people in the state of Texas. There are cases, however, when a sampling frame is not feasible. Researchers have developed different ways to sample from a population even if a sampling frame is not possible. Those types of sampling are discussed throughout this chapter.

The first step toward finding the best sampling technique that will fit your specific study is to determine the population of interest. What is your sampling frame? Maybe your sampling frame is easy to access and you can write down the names of people who make up your population or you are able to find such a list. For example, if your population of interest is students of a specific major or students of a specific university, you could visit the registrar's office and acquire the sampling frame of interest.

On the other hand, your population of interest may be working mothers. In this case, you can write down "women with children who also work." Once that sentence is in front of you, it will come naturally to ask yourself: "Is there an age group you are more interested in studying? Are you interested in mothers of small children younger than 5 or mothers of teens? Are you thinking of mothers who work full time or part time? Does it count if mothers work from home? Are you thinking of mothers in your own town or the adjacent city?" Questions like these will help you come up with a clear idea of your population and will pave the way for selecting the most useful sampling technique for your study.

PROBABILITY AND NON-PROBABILITY SAMPLING

Sampling types can be divided into two large groups: **probability** and **non-probability sampling**. The main distinction between these two groups depends on the random selection of participants and the use of a sampling frame. In probability sampling types, participants are randomly selected and a sampling frame is often used. In non-probability sampling types, participants are often not randomly selected and a sampling frame may not be available. Randomization for sampling has the same meaning as mentioned in Chapter 4 when we talked about random designs: Everyone in the population has an equal chance of participating in the study.

The type of study we are conducting determines whether probability or non-probability sampling will be employed. For example, a research question might

Everyone in the population has an equal chance of being selected.

focus on capturing human immunodeficiency virus (HIV) risk behaviors among young adults (between 18 and 24 years old). According to the Centers for Disease Control and Prevention (Patel et al., 2014), common risk behaviors for HIV include vaginal, anal, and oral sex and substance abuse. For a study to capture the frequency of these behaviors among young adults, probability sampling is desired because all members of this population (i.e., young adults from 18 to 24 years old) should have an equal chance of participating. This will result in a higher number of participants in the study.

On the other hand, you may be interested in understanding why new immigrants engage in HIV risk behaviors. You might ask, for example, whether they feel vulnerable or if they are experiencing cultural shock. Probability sampling would mean that everyone in the population has an equal chance of participating in the study, but in this scenario, the researcher may have a hard time recruiting participants. Newly arrived immigrants are still adjusting to a new culture and may even be in the process of legalization, a fact that may make their willingness to participate in a study focused on HIV risk behaviors quite low. They may be suspicious of the researcher and even mistake the scientific research with the legalization process—even more reasons to avoid participating altogether. Being able to recruit participants in this case could be difficult by probability sampling. These types of studies would benefit from non-probability sampling because of the sensitivities involved. Since only a few people may be willing to share such personal information, the researcher would first need to establish trusting relationships with participants.

We often use probability sampling in quantitative studies and non-probability sampling in qualitative studies. Both types of sampling are useful, depending on the needs and nature of the study. Some questions can help you clarify whether you need to use probability or non-probability sampling by helping you consider your study's focus and the accessibility of the population of interest. Is your population easy to reach or will it be difficult to find people who fit your study requirements? Is the topic sensitive, which may cause participants to be reluctant to talk to you? The answers to these questions will help you decide between probability and non-probability sampling first before moving to the specific types of sampling for each group.

TYPES OF NON-PROBABILITY SAMPLING

Non-probability sampling includes various techniques that do not follow the basic rule of probability sampling: the equal chance of participation. Therefore, when some people have a higher chance of being selected for a study than others, we use ***non-probability sampling.***

These types of studies often involve populations that are difficult to reach or where a sampling frame of the population does not exist. Let us take a look at some common types of non-probability sampling techniques.

Convenience Sampling

Convenience sampling is a technique that allows the researcher to select any participants who are available to participate in a study, even if they are not representative of a population. Convenience sampling often means gathering data or information on participants who are available to you as a researcher at a given point in time (e.g., people in your network or people who happen to be together with the researcher in some event). In other words, their participation happens by availability and accident. Because participation occurs by accident, convenience sampling is also called *accidental sampling*. It proves particularly cost- and time-effective in the case of **pilot studies**. Pilot studies are those initial inquiries that pave the way for larger studies. They are widely used for qualitative and quantitative inquiries. Sometimes, a research idea may sound terrific on paper, but can turn into a fiasco when put in motion. To avoid badly conducted studies, most researchers first conduct a mini-study that mimics the larger study.

For example, imagine that a researcher is planning to collect data on rape perceptions and risk behaviors of college students from five universities in a specific region. This researcher has reviewed all available studies on rape perception among college students and has created what she thinks is a perfect survey. Before a random selection of classes from the five universities at which the data will be collected, the researcher asks a friend to first allow data collection in her small classroom of 12 students. This is a convenience

FIGURE 6.1 ■ Convenience Sampling

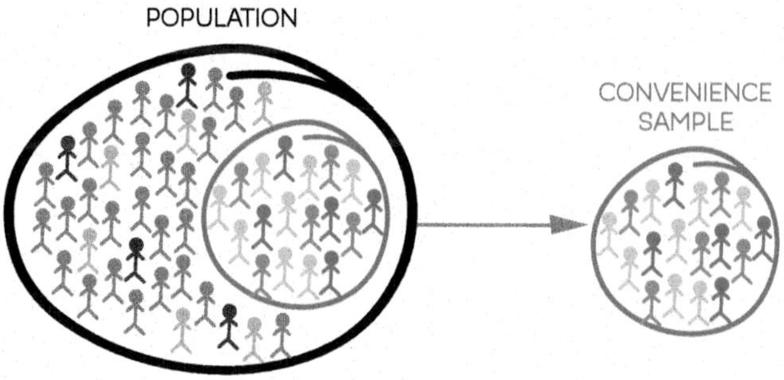

POPULATION

CONVENIENCE SAMPLE

sample; the 12 students may not be representative of the population of interest, but they are available to participate in the study.

This mini-study allows the researcher to calculate how much time it will take students to complete the surveys while evaluating the viability of the questions. It is a chance to adjust language and check for redundancy. Pilot studies are a wonderful way to test an idea before it evolves into a large-scale research study, with the added benefit of an easily accomplished convenience sample.

Pilot studies are a cost-effective and attractive way to test aspects of a study. It is important to conduct the pilot study in a fashion identical to the main study, with the added benefit of asking the participants for feedback. Were there ambiguous questions? Were there questions that seemed to be interpreted differently from what you intended? Was the questionnaire or interview a burden for participants? Did they feel uncomfortable at any time during the study? Did they have difficulties in answering a specific question? Were the scales (if using questionnaires) clear? These follow-up questions are extremely important to create a much better instrument or prepare yourself for a better interview when collecting the data. If the pilot study reveals problems with your questionnaire or the way you have framed your interview questions, you may want to consider conducting a second pilot study to test the improved instrument.

Snowball Sampling

Snowball sampling is a technique in which participants are selected by word of mouth. The researcher is able to connect with one participant and that participant finds another willing participant. The second participant finds the third participant, and so on. As is true in convenience sampling, snowball sampling may not be representative of the entire population because the researcher selects only participants who know one another and are related in some way. They may share characteristics, such as social class, income, ethnicity, sexual orientation, or religion.

Although the lack of representativeness of the entire population is a disadvantage, snowball sampling is useful when the population is hard to reach. For instance, a researcher may be conducting a study on illegal drug trafficking or prostitution, where participants will not readily come forward, regardless of the assurance of confidentiality. Snowball sampling also works well when the population of interest requires a more personal, internal connection to the study.

If snowball sampling is the technique that best fits your study, you must find one or a few people from your population of interest. These may be people you know personally, friends of a friend, or other types of acquaintances. Although snowball sampling is considered non-representative of the entire population, you should try

FIGURE 6.2 ■ Snowball Sampling

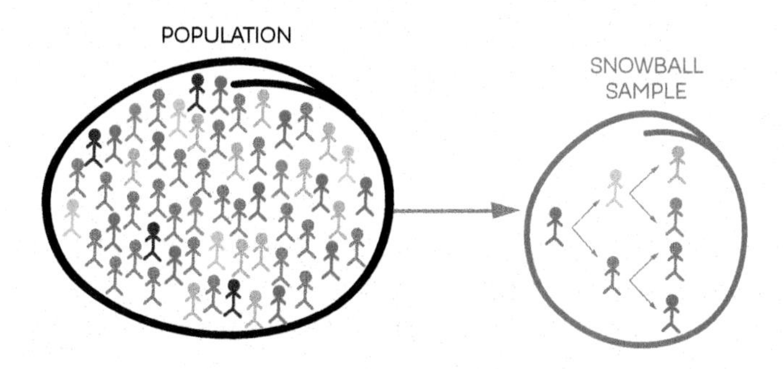

to introduce as much participant variety as possible in your sample. For example, if you have interviewed a good number of boys, but not many girls, you can encourage your next participant to recommend a girl to participate. This can be applied to any characteristic of interest.

When conducting snowball sampling, it is of utmost importance to gain the trust of the first few participants. If you can convince your participants that the study will not harm them in any way and that their identity is protected, then you have a much higher chance that they will recommend others. It may also help if the first participants have a pleasant experience overall, because they will be more likely to recommend your study to their acquaintances.

ETHICAL CONSIDERATION 6.1

TO REMEMBER WHEN SAMPLING

Ethical considerations, such as protecting participants from any potential harm or protecting their identity, are at the core of any study, using any kind of sampling. However, when using non-probability sampling, there may be a higher risk that participants' identities will be revealed because of the small sample size and the fact that they may recognize each other.

In snowball sampling, for example, one participant may be the link to the other participant, so participants are aware of who took part in the study. To protect their identities, the researcher must show participants the conclusions of the study so they are aware of how much personal information is revealed and are comfortable with that particular level of information becoming public. If participants ask that some identifiers or personal information be removed, the researcher is obligated to remove them.

Purposive Sampling

Purposive sampling is often called *judgmental sampling* and is a technique that allows the researcher to select the participants of interest for the study. Sometimes, we conduct unique studies that focus on a very specific population with unique characteristics. For example, a researcher may be interested in professional writers who were homeschooled as children. This is a very narrow population and is difficult to uncover. A skilled researcher can use purposive sampling to identify potential participants and approach them individually by using specific sampling techniques.

Homogenous sampling, for example, is a type of purposive sampling technique in which participants are chosen based on a trait or characteristic of interest to the researcher. This sample is homogenous because each participant must have the specific characteristic the researcher is looking for. One example of this sampling technique may be a researcher who is investigating the academic performance of female hockey player college students. To participate in this study, one needs to be (1) female, (2) a hockey player, and (3) a college student. The sample is homogenous in these three characteristics, which are the focus of the study.

Another type of purposive sampling is **deviant case sampling**. This type of sampling focuses on unusual or very specific cases. For example, a researcher in education may be interested in exploring why some students drop out of high school in their first year. The majority of young adolescents who are enrolled in their first year of high school choose to continue their education. The researcher is focused on only the outliers (in this case, high school dropouts). There are only a small number of outliers who would comprise this group, hence the name *deviant* case sampling.

Purposive sampling allows a researcher to handpick participants according to the characteristic under study. That benefit, however, comes at the cost of not being able to

FIGURE 6.3 ■ Purposive Sampling

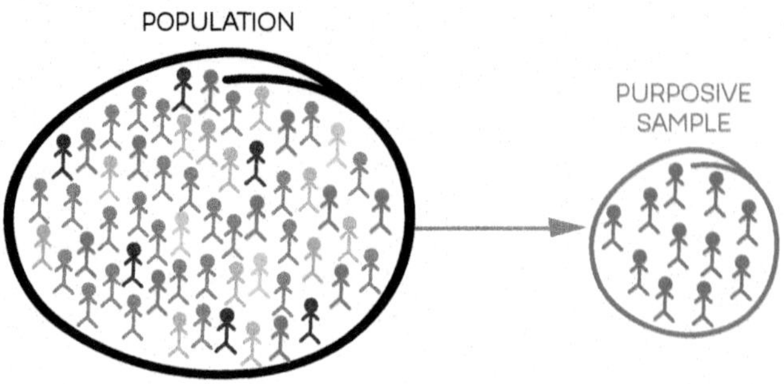

POPULATION

PURPOSIVE SAMPLE

generalize the results to the greater population and offering very low external validity. It is important to have logical reasoning in choosing this sampling choice, which is supported by the literature. In cases where the only way to explore a phenomenon is through purposive sampling, researchers can put forth groundbreaking work.

Quota Sampling

Quota sampling is a technique that allows us to compare different groups within the population of interest. To achieve the goal of comparing different groups, researchers need a sample of predetermined quota or proportions. There are two types of quota sampling: **proportional quota sampling** and **non-proportional quota sampling**.

Proportional quota sampling refers to the sample's representation of the same proportion as it exists in the entire population of interest. For example, let us assume that we are studying learning disabilities among the population of a high school where the proportion of students with learning disabilities makes up 20% of the entire student body. If we are using proportional quota sampling, our sample should also represent this same 20% proportion of students with learning disabilities. In practice, this means that we can only collect information from participants who fit into the predetermined quotas, so in other words 2 out of 10 students in our study should have a learning disability so that we fill up the proportional quota we need.

Non-proportional quota sampling, on the other hand, uses a different quota from the one found in the population of interest because the study's aim is to compare two or more different groups of interest. For example, if we were to compare the academic achievement of students with learning disabilities compared to the rest of the student

FIGURE 6.4 ■ Proportional Quota Sampling

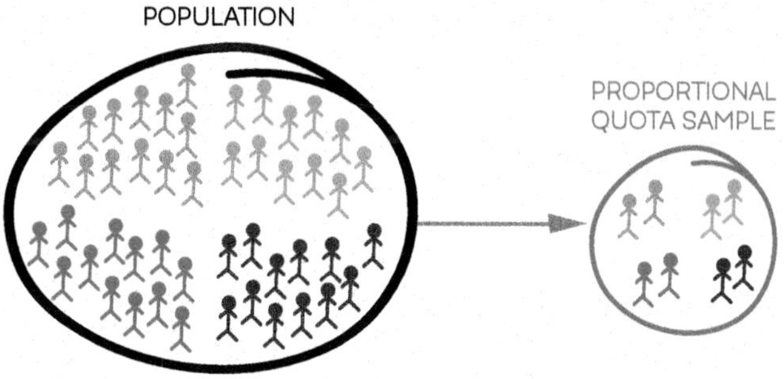

FIGURE 6.5 ■ Non-Proportional Quota Sampling

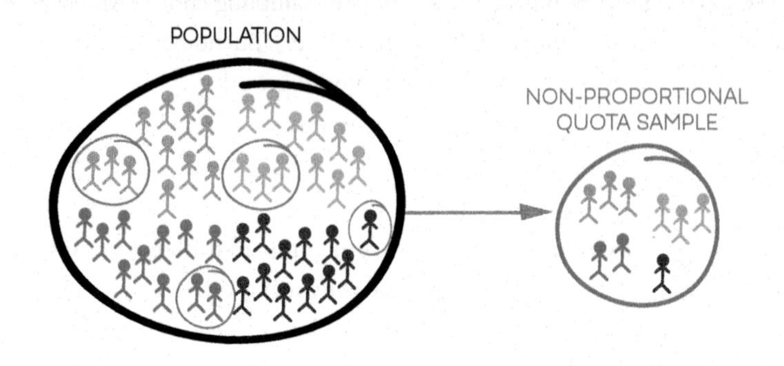

body, it may be best for our sample to have a proportion of 50:50 rather than 20:80. In this case, we are not interested in the actual proportion of students with learning disabilities in the population, but we are more concerned with our own predetermined quotas because they best fit the comparison we are trying to analyze.

RESEARCH WORKSHOP 6.1

TIPS TO REMEMBER WHEN SELECTING NON-PROBABILITY SAMPLING

Here are a few rules of thumb to keep in mind while considering a non-probability sampling technique:

- Non-probability sampling designs are often viewed as cost-effective. This may or may not be the case if one factors in the time we need to spend conducting the study compared to probability designs.
- Non-probability designs are an excellent choice for exploratory, small studies where we are just testing the waters in an area that not much research has been conducted.
- Generalization will be difficult—we may not quite be able to generalize our findings to a larger segment of population, and that is a problem to consider.

TYPES OF PROBABILITY SAMPLING

There are many types of probability sampling, but the most commonly used are simple random sampling, stratified random sampling, cluster random sampling, systematic random sampling, and multistage sampling. The shared feature among these types of sampling is randomization, but they vary in the methodology used to choose participants. Deciding on the most appropriate type of sampling for a study depends on its aim.

Simple Random Sampling

Simple random sampling is a sampling procedure that relies on complete randomization without any specific boundaries. If we definitively know the population to investigate, this sampling method is preferred. Simple random sampling is conducted in one step using techniques such as random number generation, picking numbers/participants from a fish bowl, or other simple techniques. Random number generation is a simple computational technique that aims to generate a set of numbers that are truly random and not related to one another in any other way. Statisticians have created a series of computations that leads to a set of numbers generated this way—just as if we were to roll the dice every single time.

The two biggest advantages of this method are that every member has an equal chance of participating and it ensures representation of the population. Its biggest disadvantage is that it *requires* a complete list of the population from which the sample is drawn. For example, if our population of interest is all the pediatricians in a town, we can rely on a phone book, ask the town hospital for a complete list, or find a comprehensive list of pediatricians' offices and their contact information online. In this case, we can use simple

FIGURE 6.6 ■ Simple Random Sampling

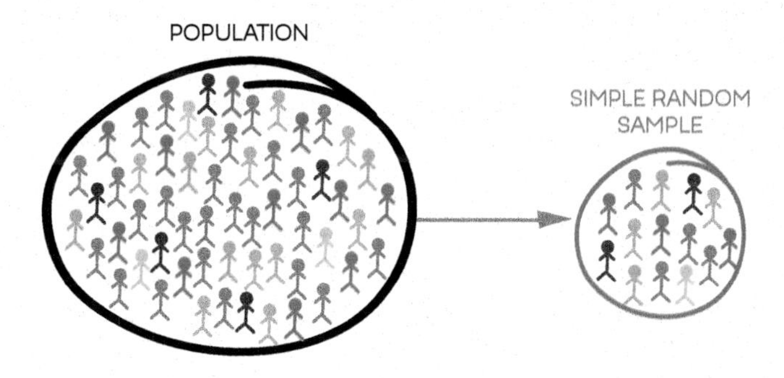

random sampling because everyone in our population of the town's pediatricians has an equal chance of participating in our study. For other populations, it can be arduous or impossible to achieve a compilation of all the members.

What if our population is the homeless population of this town? We may be able to find a list at the local shelter or soup kitchen, but, even if such a list is available, it will be incomplete. There are many homeless people who do not show up at a shelter or who go there sporadically. In addition, this population will likely not have ready contact information, and even with a list, it will be hard to make contact. In this case, simple random sampling is impossible, so other sampling techniques are needed.

Now may be a good time to ask whether simple random sampling is the right method for your study. Can you create a complete list or a sampling frame of the participants your study focuses on? If the answer is yes, then you will be able to use simple random sampling. Finding available lists may require some work, but it increases the external validity of the study. Some ideas for creating such lists may be using zip codes, house numbers, listed phone numbers, hospitals, doctor's offices, registrar's offices, police records, or any other types of lists that are publicly collected from large institutions.

Stratified Random Sampling

Stratified random sampling is a technique that becomes valuable when the study is focused on understanding, comparing, or analyzing different groups of a population. This requires equal numbers of participants from each group, for example, men and women, different ethnicities, marital status, and so on. The researcher creates lists of different groups called **strata**. Once it is clear that the study is focused on two or more strata, the researcher may randomly choose an equal number of participants from each group. When the required number is attained for one strata, the researcher stops adding participants for that group.

Let us look at an example of stratified random sampling. Imagine that a researcher is interested in understanding whether young women in college were more or less likely to engage in illegal substance abuse than young men in college. This researcher plans to gather data from one specific college, so it is a relatively easy task to have a complete list of females and males enrolled in this college. Once the lists are available, there are two possible ways of selecting the sample: (1) the researcher randomly picks the same number of female participants and male participants from a list of females and a list of males, or (2) the researcher may use the complete list of all students enrolled in this college and randomly pick participants—if a female is selected, that name is entered into the strata for females, and if a male is selected, that student's name is put into the strata for males.

If the researcher wishes to collect data from 100 participants where the gender division is 50:50, then once one of the strata has reached the maximum, the researcher stops looking for additional participants for that group. Therefore, if we selected 50 females for this study, but we have only 35 males, if we pick another female, we do not select that participant and continue with random selection for males only. The procedure continues until the same number is reached for both groups.

Stratified random sampling allows for the comparison between groups with random sampling. When appropriate, we may alternatively use **proportionate stratified sampling** or **disproportionate stratified sampling**.

Proportionate stratified sampling is the type of sampling that follows the proportions of the population, but also creates specific strata that are of interest to the study. For example, say a researcher is interested in investigating the relationship between parents' education and teenagers' illegal substance abuse and hypothesizes that children of parents with a bachelor's degree or higher are less likely to abuse illegal substances. The researcher looks at U.S. Census Bureau data to select a sample and finds that in 2015, 33.3% of the U.S. population had a bachelor's degree or higher (Ryan & Bauman, 2016). To accurately reflect the wider population, the researcher employs proportionate stratified random sampling, selecting 100 teenagers for this study, of which 30 to 33 have at least one parent with a bachelor's degree or higher, in keeping with the U.S. Census Bureau's statistic of 33.3%. Although the selection of the participants is random, the researcher stops collecting data from teenagers whose parents have a bachelor's degree or higher once 30 to 33 such cases are collected. By using proportionate stratified sampling, this study sample represents the entire U.S. population.

Disproportionate stratified sampling is a sample in which the proportions are not equivalent to the proportions in the entire population. An example of disproportionate stratified sampling would be a study on the value of family ties and family connections between African Americans, White Americans, and Hispanic Americans. For this type of comparison, it is important to have equal groups of African Americans,

FIGURE 6.7 ■ Proportionate Stratified Sampling

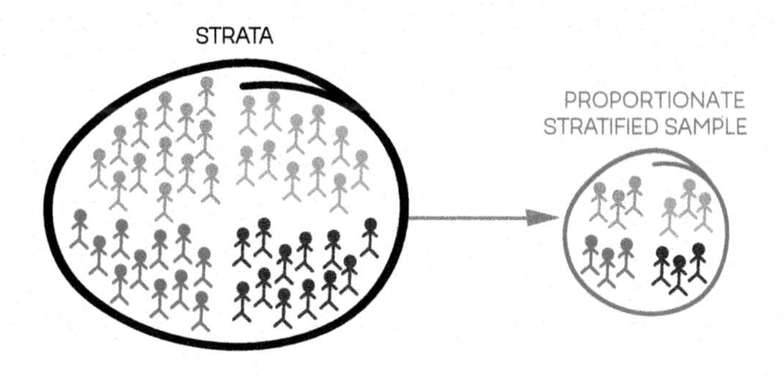

RESEARCH IN ACTION 6.1

TWO STUDIES USING PROPORTIONATE AND DISPROPORTIONATE STRATIFIED SAMPLING

Source: Cohen, Dillon, Gladwin, & De La Rosa (2013). American parents' willingness to prescribe psychoactive drugs to children: A test of cultural mediators. *Social Psychiatry Epidemiology, 48*, 1873–1887.

STUDY EXCERPT

Data were collected by trained bilingual interviewers under the supervision of Florida International Institute for Public Opinion Research from May 11 until October 8, 2009. . . . In all, 35,311 different telephone numbers were called yielding 1,145 complete interviews. Selecting quotas of respondents who met the inclusion criteria was accomplished in three phases: (1) filtering out-of-scope cases, (2) screening the remaining (eligible) cases to identify qualifying households, and (3) completing a full interview once a qualifying household was identified.

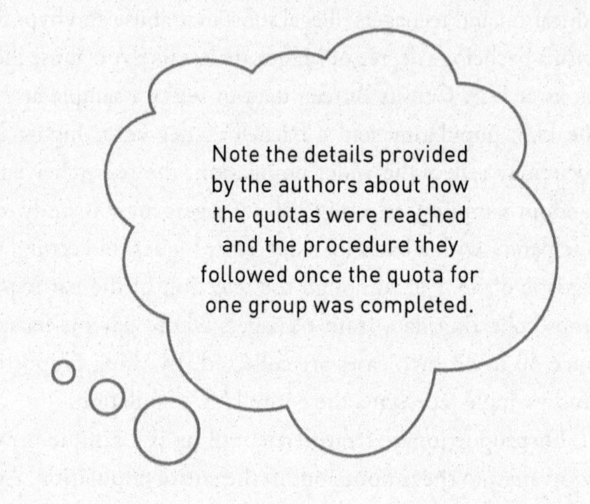

PROPORTIONATE STRATIFIED SAMPLING

Cheng, Chen, Yen, & Huang (2012). Factors affecting occupational exposure to needlestick and sharp injuries among dentists in Taiwan: A nationwide survey. *PloS ONE, 7*(4).

STUDY EXCERPT

We used a multistage proportional stratified sampling method to obtain a nationally representative sample. According to a government urbanization index, all 359 townships in

Taiwan were divided into 2 central cities, 3 provincial cities, and 16 counties. We defined our research areas as 2 central cities, 3 provincial cities, and 8 counties, which were randomly selected from 16 counties. Then, a total of 60 hospitals and 340 clinics were randomly selected from the 13 research areas when a random number generated by a random number generator was less than the probability proportional to size sampling method.

See how the proportional stratified sampling differs from the first example shown here. In this case, areas are first defined, cities in those areas, and finally counties. From the 16 counties, the researchers noted that there are 60 hospitals and 340 clinics. These were the last to be randomly selected. The procedure is quite different from the first example where the quotas were pre-defined regardless of the population.

White Americans, and Hispanic Americans in the sample, even though these equal groups do not proportionally reflect the population-at-large. Through a random sampling procedure, the researcher can achieve a sample where 30 participants are African Americans, 30 participants are White Americans, and 30 participants are Hispanic Americans. Once a category is filled, the researcher stops collecting participants from that group.

FIGURE 6.8 ■ Disproportionate Stratified Sampling

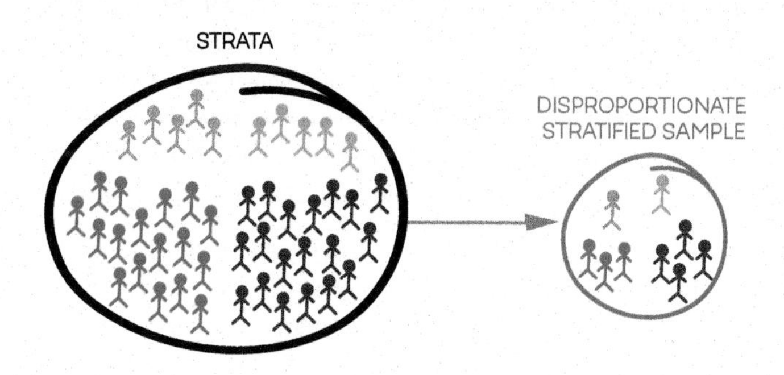

RESEARCH WORKSHOP 6.2

TIPS TO REMEMBER WHEN SELECTING A PROBABILITY SAMPLING METHOD

Probability samplings are quite attractive to many researchers for a variety of reasons, but perhaps the most important is its capability of generalizing the findings to a larger population. However, there are a few things to keep in mind before selecting probability sampling:

- The study will likely be costly in order to be time-effective, or it will be very labor- and time-consuming with the possibility of lowering some costs.
- Probability sampling techniques are truly great for conducting larger studies, reaching out to many participants, and testing solid hypotheses.
- If the study you are conducting is exploratory in nature, still unexplored from the literature, or you may not be completely certain about the hypotheses, questions you are asking, and any other details, probability sampling may not be the best option.
- Once the data are collected, it will be impossible to go back to participants and clarify something that you may have not quite captured in the first round of questions—so the questions need to be truly well thought out and able to encapsulate the constructs you are trying to capture.
- Probability sampling techniques are often easier to use if you have already conducted a study in the past and this is not your first attempt at conducting research.

Summary

Sampling is the procedure researchers use to select a sample of (a few) participants for their study that is as representative as possible of the entire population. A population is the entire group of people the researcher is interested in studying (i.e., if the study focuses on adolescents, all adolescents are the population of this study). A sample is a smaller group of people taken from the population of interest. Researchers use samples to analyze and draw conclusions that may be later generalized to the entire population. To do so, researchers use a sampling frame. A sampling frame is a complete list of the entire population of interest.

There are two main types of sampling: probability and non-probability sampling. The main difference between these two types of sampling is randomization, or whether the participants were randomly selected. When participants are randomly selected to participate in a study, researchers are using probability sampling. When other ways are employed to select participants, researchers are using non-probability sampling. Some types of probability sampling are simple random sampling, stratified random sampling, cluster random sampling, systematic random sampling, and multistage sampling.

Simple random sampling is a sampling procedure using one-step random techniques, such as using a computer number generator. To conduct simple random sampling, the researcher needs a complete list of the population. Stratified random sampling is a sampling technique that allows the researcher to differentiate between two or more groups of interest, called strata. Therefore the researcher is able to still use random selection, but still have the same number of participants for each strata of interest. Stratified random sampling can be either proportionate stratified random sampling or disproportionate stratified random sampling. Proportionate stratified random sampling is a technique that suits a study focused on mimicking the proportions of the population regarding one or more characteristics. These types of studies manage to use random sampling and still preserve the same proportions found in the entire population on one or more specific characteristics. Disproportionate stratified random sampling is a technique that suits studies focused on conducting comparisons between groups of the population. These studies are interested in having a specific proportion of the groups of interest that may not translate in the same proportion in the entire population. Cluster random sampling is a technique of sampling that allows the researcher to create separate random clusters first and then randomly select participants from each cluster. A cluster is a group of the population within which the participants are going to be randomly selected. A cluster needs to be as similar to the entire population as possible. Systematic random sampling is a sampling technique based on selecting the *n*th number on the sampling frame where *n* is the sampling interval. The sampling interval is the ratio between the sampling frame and the sample size. Multistage sampling is a sampling technique based on the combination of two or more types of sampling together to best suit a specific study.

Non-probability sampling offers various types of sampling techniques to select participants who may not be representative of the entire population and are not randomly selected. Some types of non-probability sampling are convenience or accidental sampling, snowball sampling, and purposive sampling. Convenience or accidental sampling is a technique that relies on the participants who are available and willing to take part in the study. Convenience sampling is ideal for pilot studies. Pilot studies are mini-studies that test the quality of the research designed or other characteristics of a larger study. Snowball sampling is a technique that relies on initial participants finding other people who are willing to participate. The word of mouth from one person to another makes these studies very suitable for research focused on vulnerable or hard-to-find populations. Purposive sampling is a technique that allows researchers to individually pick participants for their study. Commonly the goal of such studies relates to some very specific characteristic that is difficult to find in the general population.

Some other characteristics of sampling are sampling error, confidence interval, and saturation. Sampling error is the difference between the sample used in our study and the entire population of interest. Being able to estimate the sampling error helps understand more about the population. Confidence interval shows the percentage of confidence the researcher is aiming for in the study. The margin of error represents the acceptable percentage of error in a study. Saturation determines the sample size for non-probability studies and relates to the moment when the researcher is not getting any additional new information from the participants. Once the researcher reaches the saturation point, he or she has also established the sample size for that study.

Key Terms

Convenience sampling: a technique that allows the researcher to select any participants who are available to participate in a study, even if they are not representative of a population.

Deviant case sampling: a type of purposive sampling that focuses on unusual or very specific cases.

Disproportionate stratified sampling: a type of sampling in which the proportions are not equivalent to the proportions in the entire population, but specific strata are created.

Homogenous sampling: a type of purposive sampling technique in which participants are chosen based on a trait or characteristic of interest to the researcher.

Non-probability sampling: a sampling type in which participants are often not randomly selected and a sampling frame may not be available.

Non-proportional quota sampling: a type of quota sampling that uses a different quota from the one found in the population of interest because the study's aim is to compare two or more different groups of interest.

Pilot study: a mini-study that mimics a larger study.

Population: the entire group of people that are the focus of the study.

Probability sampling: a sampling type in which participants are randomly selected and a sampling frame is often used.

Proportional quota sampling: a type of quota sampling that refers to the sample's representation of the same proportion as it exists in the entire population of interest.

Proportionate stratified sampling: the type of sampling that follows the proportions of the population, but also creates specific strata that serve the focus of the study.

Purposive sampling: also called judgmental sampling; a technique that allows the researcher to select the participants of interest for the study.

Quota sampling: a technique that allows a researcher to compare different groups within the population of interest.

Sample: a subset of the population.

Sampling: the procedure used by researchers to select a subset of the population that can be used to conduct a scientific study.

Sampling frame: a list of the entire population of a study.

Simple random sampling: a sampling procedure that relies on complete randomization without any specific boundaries.

Snowball sampling: a technique in which participants are selected by word of mouth.

Strata: a list of different groups of people that requires an equal number of participants from each group.

Stratified random sampling: a technique used when a study is focused on understanding, comparing, or analyzing different groups of a population.

Taking a Step Further

1. What divides probability sampling from non-probability sampling?
2. How does the sample relate to the population?
3. When do we need to use snowball sampling and what is a major drawback of it?
4. What is the procedure for stratified random sampling?
5. Can you think of an example that calls for homogeneous sampling?
6. How do we know that we have reached saturation?

7

DATA COLLECTION FOR QUANTITATIVE RESEARCH

CHAPTER OUTLINE

WHAT WILL YOU LEARN TO DO?

1. Review the characteristics of experimental research
2. Describe the steps necessary in preparing and collecting data using a questionnaire
3. Compare and contrast the different methods of data collection

EXPERIMENTAL RESEARCH

Recall from Chapter 4, that quantitative studies include very little if any context or textual information from the data collected. These studies often standardize the retrieval of information from participants in the form of numbers, and even express qualitative information like race, gender, religiosity, and others numerically. The data for quantitative studies are often collected by a computer or additional team members, so there is little if any relationship or contact between the researcher and the participants. Most importantly, quantitative research is optimal for large-scale studies and reaching out to large samples of population, but the information collected is often surface information rather than deep or comprehensive. You also remember that **experimental research** refers to conducting an experiment or intervention on one of the randomly assigned groups in the study while we control our findings with a second group. In an ideal experimental design, participants are randomly assigned into two different groups: the experimental and the control group. The experimental group is exposed to a variable of study (e.g., testing a new drug or intervention training), whereas the control group is kept in the same conditions as the experimental group with the exception of not undergoing any manipulation. The results are collected from both groups to identify the differences.

From childhood we are aware of the joy of conducting an experiment, but we also know, even then, about the amount of work, details, and patience needed to conduct a successful experiment.

Conducting an experiment is exciting and fulfilling, though it takes a lot of effort. Experiments require close attention to detail and, even so, many aspects may not go as planned. The researcher needs to complete a few basic requirements:

a. Randomly assign participants into control and experimental groups

b. Conduct a pretest and a posttest to measure the expected areas of change

c. Protect findings from other external explanations

A straightforward experiment may, for instance, investigate changes in students' perceptions of poverty after participating in a poverty-related simulation, as compared to a control group who watched a romantic movie instead. The alternative hypothesis may predict a relationship between participating in a poverty simulation game and increased awareness of the difficulties of poverty. The null hypothesis would predict that the perception of poverty would not be affected by playing a simulation game. The researcher would look at the differences in perception between the groups and can either reject or fail to reject the null hypothesis based on the outcomes. The information collected from the pretest and the posttest from both groups comprise the data. The data are handled carefully, entered into the appropriate software, cleaned, and analyzed. Care in executing the myriad details of the study and anticipating anything that might go awry will make for a successful experiment.

RESEARCH WORKSHOP 7.1

CONDUCTING THE POVERTY-SIMULATION EXPERIMENT

If you are interested in conducting an experiment, following the steps for the poverty-simulation example should work for any small-scale experiment. The experiment would proceed as follows:

1. Two groups of students of equal size (e.g., 10–15 students) should be randomly selected. You must balance the makeup of participants in terms of race, gender, age, income, marital status, or any other characteristics that may potentially interfere with this experiment.
2. Both groups are given a pretest on their perceptions about poverty. You can find a variety of pretests online on any topic that measures attitudes or beliefs of people.
3. One group (the experimental group) plays a poverty simulation game for a few minutes, in which they experience what it means to be without any resources and are incapable of knowing how to navigate the system. Such scenarios are available on several webpages, such as http://playspent.org/.
4. The control group is given a romantic movie to watch, preferably a popular one that most students have already seen.
5. Both groups are given the posttest on perceptions about poverty, which has the same questions as the pretest.
6. The researcher then measures the differences in attitudes between the pretest and posttest for both groups.

Experimental studies like this allow us to examine outcomes that occur when different variables are manipulated. We must question whether the changes in the experimental group are due to the intervention and whether we can predict the outcome of a specific behavior. The results of our experiments empower us to push forward in new directions.

ETHICAL CONSIDERATION 7.1

CODE OF RESEARCH ETHICS

When conducting experimental research, researchers must be vigilant about possible ethical issues. Participants must be treated with respect and their needs considered. There should never be an instance when a participant is harmed or put at risk. If risks are inherent in a study, the participants need to be formally notified of these risks and provide consent to participate. Also, fair and consistent treatment of participants must be kept in mind. Differential treatment can put a study at risk, as it violates the norms of research ethics. To familiarize yourself with the entire code of research ethics, please go to https://www.citiprogram.org/index.cfm?pageID=265 to see what constitutes responsible research conduct. Familiarity with these ethical rules is necessary before any study is launched.

QUASI-EXPERIMENTAL RESEARCH

We may find ourselves in a position when a classic experimental design is not possible for various reasons. Some possibilities include our inability to randomly assign participants to control and experimental groups; it may be impossible to collect pretest data, or we may only have non-equivalent groups available. So researchers often find themselves in the position where quasi-experimental research is the only possible choice. Quasi-experimental studies have the great benefit of still allowing the researcher to conduct some form of intervention, training, or other experiment, without the concerns of not fulfilling all the required conditions for the classic experimental study. As detailed in Chapter 4, there are a number of quasi-experimental designs available for researchers, such as *randomized one-group posttest-only design*, *randomized posttest-only control group design*, *non-random posttest-only control group design*, *non-random pretest posttest control group design*, and *non-random one-group pretest posttest design*. Because of the criteria of conducting an experimental design and other restrictions we may face, such as the impossibility of randomization, lack of time, lack of financial resources, and others, we often consider quasi-experimental studies since they still have the potential to measure the impact of some intervention or experiment in a group of participants.

For example, an educational researcher may be interested in how young children learn and has developed a theory that statistics may be best introduced to children at the same time that they are learning basic mathematics, during the first grade. The researcher may not be able to conduct a pretest or even have the ability to have a control group. However, the researcher can randomly select an elementary school from the schools in the area where the study is conducted and introduce basic statistics to all the first graders in the school. There will be no control group and no pretest, but the researcher will collect posttest data once the academic year of introducing basic statistics ends. Additionally, the researcher randomly selected this specific school, so she or he was able to conduct a randomized one-group posttest-only design.

Following this same example, we can imagine a different situation where the researcher who is interested in introducing basic statistics to first graders believes that if students are introduced to statistics concepts this early, they may be able to absorb mathematical skills more quickly and perform better. In this slightly different research scenario, the researcher is still randomly selecting a school in the area, but this time considering schools where there are two or more groups of first graders. In this case, it may be simple to randomly select one class of first graders who will be introduced to statistical concepts in addition to mathematics and one group of first graders who will continue their mathematics program without any introduction to statistics. It may still be impossible to conduct a pretest, but the researcher will collect posttest data from both groups of first graders to see if there are differences in their mathematical skills. So the researcher has conducted a randomized posttest-only control group research. As you can tell from these examples, quasi-experimental research is friendlier to adjust to the specific circumstances of the research study we want to conduct and may be cost- and time-effective to the researcher.

PREPARING AND COLLECTING DATA THROUGH QUESTIONNAIRES

Conducting surveys is a common procedure for data collection used in various fields of study, but especially in the social sciences. Although the terms *survey* and *questionnaire* are often used interchangeably, they mean different things. A **survey** refers to the method of data collection, whereas a **questionnaire** is the instrument containing the questions. We must determine whether the questionnaire has been tested for reliability and validity. If the questionnaire was never tested and we know little of whether it is a reliable and valid instrument for data collection, we refer to it as a **non-standardized survey questionnaire**. Alternatively, if we are using a questionnaire that has been tested and is a reliable and valid form of data collection, we refer to it as a **standardized survey**

Social sciences are built around people's behaviors, thoughts, opinions, feelings, and preferences. Surveys are quite practical and useful in capturing a glimpse of those.

questionnaire. To remember this difference, we can think of the standardized tests widely used in our schools to measure our mathematical, writing, or reading skills.

These tests are *standardized* because they have been tested many times before, and researchers believe that these tests will repeatedly result in the same outcomes. In other words, they have high validity and high reliability. Non-standardized survey questionnaires often measure a variable that may not have been tested before or that has been tested differently. In this case, we have no indication of the questionnaire's reliability or validity. For these instruments, we often say that we are taking the results at face validity, meaning that we will consider and analyze the data as if they were valid and reliable. In other words, we are saying that if we consider these data to be an accurate and a true representation, then our analysis shows a particular relationship between variables. By claiming face validity of the non-standardized instruments, we are still able to analyze the data and bring forth new knowledge, but we remain aware that our data collection instrument was not tested for reliability and validity.

Designing questionnaires is a skill that takes years to develop. This skill should be a source of accomplishment and pride and becomes a desirable resume builder. Over time, researchers learn to use a unique skill set to construct effective research questionnaires. There are nine steps to follow in creating questionnaires.

Step 1. Develop a list of constructs.

Developing a complete list of all the constructs to be measured in a questionnaire stimulates thinking about ways to measure them. First, looking at your hypotheses helps begin this list. For instance, if we hypothesize that gender influences college students' choice of major, then gender is one construct and the choice of major is another. However, do we also want to control for race, ethnicity, age, income, marital status, parental education, religiosity, and employment status? All these controllers are also constructs to measure. Our hypothetical list includes the following:

- gender
- choice of major

- race
- ethnicity
- age
- income
- marital status
- parental education
- religiosity
- employment status

Just by compiling this list, we begin to wonder about collecting additional information. Not only do we wish to inquire about the participants' majors, but also we want to ask whether they are in the major they wished they were in. Is it possible that some college students are in the major of their parents' choosing? How about inquiring about that angle? What if we looked at their minors? Here is our revised list of constructions (the italicized words are additions to the previous list):

- gender
- choice of major
- *major the participant wished he or she had*
- *choice of minor*
- *reasons why the participant is in this major*
- race
- ethnicity
- age
- income
- marital status
- parental education
- religiosity
- employment status

The ways we think about these constructs represents our creativity and instincts as researchers. One way to guide the process is to think about the outcome. What type of outcome can we predict? What questions can we ask at the beginning of the study rather than later? Perhaps it will be important to know what future job participants are interested in. This might explain their choice of a major. Also, are they satisfied with their major? You will want to consider various lines of questioning and their impact on research outcomes. Finding out whether participants are happy with their choice of a major may add depth to our study, along with understanding the relationship between gender and major. Here is the revised list (new information underlined):

- gender
- choice of major
- major the participant wished he or she had
- choice of minor
- reasons why the participant is in this major
- the job the participant is aiming for
- participant's level of satisfaction with his or her choice of major
- race
- ethnicity
- age
- income
- marital status
- parental education
- religiosity
- employment status

Step 2. Determine how each construct may be accurately measured in the form of a question.

Now that we have a list of constructs, how do we measure them? One possibility is to look for established survey instruments that may be related to your study. Various methods of measuring gender, race, ethnicity, age, income, marital status, parental education, and employment status are readily found in the literature. Some potential sources in which to find questions are U. S. Census data, National Institutes of Health surveys,

Centers for Disease Control and Prevention questionnaires, Department of Education questionnaires, and many others.

The more challenging constructs are the ones specific to your study. For example, how do we measure why the participant chooses a major? There are two ways to solve this problem: (1) the researcher consults the literature, looks for scientific possibilities, and enumerates all of these on the questionnaire, or (2) the researcher creates a brief, open-ended question.

Other constructs that may be open to interpretation are level of satisfaction and religiosity. Both may be measured using existing instruments. Which of the known options will suit the needs of your study? You should always select the option that will strengthen your study. For example, the level of a student's satisfaction with his or her major may be measured by a numbered scale or a set of questions. Which type of data will best serve your needs? Measures of religiosity will vary from just one question to complex instruments with 10 or more questions. If the construct of religiosity is crucial to one's goals, one would probably choose the more in-depth version. If the concept of religiosity is not as integral to the results, the researcher may choose to use just a couple questions. Decisions are also based on the time available to complete the questionnaire, the length of the questionnaire, and particularly the significance to the research questions at hand.

Step 3. Think of all possible answers to the questions.

Some of the questions will be simple, such as gender. Possible answers include (a) female, (b) male, (c) transgender or transsexual, and (d) no answer. Sometimes questions require a bit more consideration, like one's employment status. A person may be employed full-time, part-time, retired, unemployed, or self-employed. That provides the researcher with some basic information, but knowing the duration of this status adds even more understanding. Including additional responses will aid the study, for example, to know whether the participant has been unemployed less than a year or over a year. Answers should always be *mutually exclusive*. An answer is mutually exclusive when it does not overlap with another possible answer choice. For example, a participant cannot have the option of (a) unemployed, (b) unemployed more than a year, and (c) unemployed less than a year, because this does not limit the choices to just one.

Giving full consideration to all the possible questions and responses requires some exploring of the literature, along with critical thinking. For example, in exploring possible jobs the participant might be interested in, the researcher may add various professions, such as business, education, medicine, nonprofit, law, and police enforcement, but they will likely be too broad to function as closed answers. One might either create an extensive list of broad choices or allow open-ended responses.

Step 4. Avoid biased, misleading, socially desirable, or double-barreled questions and overly technical terms.

The wording of the questions needs to be clear and grammatically correct, and it needs to avoid misleading or biased language. For example, if we were to explore marital satisfaction and ask participants, "Do you have any problems with your spouse?" this would prompt thinking about problems in the relationship and subtly lead the participant to answer in a certain way. By asking, "How do you rate the relationship with your spouse?" and giving discrete choices that vary from "fantastic" to "problematic," we avoid guiding participants' answers.

Use of technical terms can confuse participants. For example, we might be trying to understand when children start showing fears and anxieties so we can capture the time when the *amygdala*—a tiny brain structure that regulates emotions—starts developing. If we are surveying parents of toddlers and ask, "When did you see the first sign of amygdala development in your child?" we may not receive an accurate response since most parents will not know what it means. Rather, if we ask them, "Do you remember the first time your child started to show any type of fear or anxiety toward something (e.g., going on an elevator)?" we would receive more accurate answers.

When it comes to wording the questions on questionnaires, researchers need to be attentive to socially desirable answers. Sociologists refer to the term **social desirability** when they try to describe study participants' tendencies to want to *look* good and avoid the truth. People may become uncomfortable when asked about situations that are perceived as negative or undesirable by society and admit only to answers that are perceived as *desirable*.

Good parenting is one such socially desirable characteristic that can be sensitive for participants. At some point, you have probably encountered a screaming toddler in a supermarket or other public place. If you recall any of these instances or pay attention the next time you see it, you will notice that the parent accompanying the child becomes very disconcerted in the situation. It is socially undesirable to have a misbehaving child in public—although it is developmentally normal for toddlers to scream when things don't go their way. To prevent the tantrum from happening again, most parenting books suggest ignoring the misbehaving child. Although that may be easily implemented at home, it is socially desirable to attend to a screaming child in public. Social desirability is very powerful, not simply for parents in public places, but also for participants in our study.

If we were to investigate how much parents read to their toddlers, we need to be careful in phrasing the question because reading to small children is a socially desirable activity. Not reading to small children may be perceived as a parenting problem in our society, so we want to be delicate in crafting the question. Since we are seeking accurate answers, we need to consider all possible problems that may arise. Consider these following questions:

A. How many books do you read to your toddler during a typical day?

B. How many books are you able to read to your toddler during a typical day?

The first question may prompt inaccurate answers because of the social desirability of reading to toddlers. We may find that parents exaggerate their answers, an undesirable outcome. Parents may think that if they do not read much to their child, they are not very good parents. The second question (B) releases the blame on the parent by adding "are *you are able to read,"* which subtly signals that if you are not reading enough to your child it is because you are unable to do so. Parents of toddlers may be more likely to more accurately answer the second question.

Double-barreled questions ask about more than one concept, but allow only one answer from the respondent. These are very confusing to participants who may feel out of place and are unsure of how to answer. Respondents may think that their specific case was not included in the answers or that the questions were not thoroughly considered. These attitudes can have negative consequences for our survey, so we should avoid double-barreled questions. A question with only no/yes options that asks, "Were you treated with respect by the nurse and the doctor at our facility?" does not consider the alternative when the nurse was very respectful to the patient, but not the doctor, or vice versa. These questions are puzzling and can lead to inaccurate answers from respondents. One simple way to avoid double-barreled questions is to make sure we are asking about only one concept and not two, or test our questions with a third party and get their feedback.

A researcher needs to consider the population that will be surveyed. If the population is a specialized one, the rule about technical language does not apply (depending on the study). Additionally, if the population does not speak English or speaks English as a second language, the questionnaire may need to be in the native language of participants. A final, important rule about language is to avoid vague expressions or wording. The questions need to be clear without double meanings or various interpretations, because that can lower reliability.

Step 5. Organize in a manner that attracts and holds participant attention.

It is often suggested that demographic questions can bore participants and change their attitude toward the entire questionnaire. Their attention may decrease as they answer each question, and they are less willing to respond accurately. A good rule of thumb is combining all the demographic questions at the end, after the attention-grabbing questions have been completed. This best practice may pertain to sensitive surveys as well, since participants may be more willing to answer accurately if they have not yet exposed their identity through demographic questions.

Another rule of organization is to follow a logical continuity. If there is a group of questions on religiosity (e.g., questions asking about the frequency of attending a religious institution, the frequency of prayer, and whether the participant believes in God), the researcher may choose to group them together. If you were to ask one question about religion, the following question asks about age, and then the third question asks about the frequency of reading to toddlers, the participant may become distracted. It is important to group all questions together logically and use subheadings, which indicate the change of focus in the questions.

On the topic of organization, researchers should be wary of designing questions that use only scale measurements. After answering a few of them in a row, participants may start skipping questions or selecting the same rating repetitively (e.g., strongly agree or very unlikely).

Step 6. Use clarity and brevity.

Researchers should make sure that all the questions on their questionnaire are clear, are simple, and lack ambiguity. This may require a bit of time and a few trials. Preparing the simplest and cleanest questions is well worth the effort and will pay off once the questionnaire is running smoothly. When questions are difficult to understand or include any unintentional double meaning, the respondents may be turned off and answer either incorrectly or not at all.

Meanwhile, constructing the shortest possible questions is as important as their simplicity. You are well aware of how your own college papers have a higher chance of success if they are clear and to the point, without unnecessary wording. The same logic applies to questionnaires. The ability to express your exact meaning with brevity requires writing skill as well as questionnaire design expertise.

Step 7. Pay attention to contingency questions.

When the questionnaire approaches final formatting, begin to anticipate follow-up questions, known as **contingency questions**. We often ask respondents about a particular issue and their answer triggers another question. This requires the questionnaire designer to think ahead. For example, imagine that we are conducting a study about the types of psychiatric medications used by college students to handle anxiety. Initially, we need to distinguish between students who use psychiatric medications and those who do not. Our first question may be "*Do you use psychiatric medications:* Yes or No?" When our participants choose "yes," we follow up to ask about the type of medications they use. If their answer is "no," we may want them to continue to the next question. When using contingency questions, it is important to have a clear progression of questions to ensure that participants will not be confused about what question to answer next. You probably have had experience with similar surveys, where you find yourself asking, "*But I already answered that I don't use psychiatric medications. Why are they asking me about which type I use?*"

Step 8. Create an answer scale.

When measuring attitudes, beliefs, and opinions, we may wish to have all the answers grouped together in a scale, so we can see how people think and feel about something. There are different types of scales: the Likert scale (likability scales that are usually constructed from very likely to very unlikely in scales of 4 or 5), agreement scales (often varies from strongly agree to strongly disagree), and other multipurpose scales (for example, the ones that include choices from extremely to not at all). Research has shown that many of these terms, such as *agree, very likely, often,* and so on are subjective and can be interpreted differently by different people (Barnes, Cerrito, & Levi, 2003; Schwartz, 2005). However, we can still examine how people measure their own opinions and beliefs in a scale from a maximum to a minimum.

It is beneficial to remember that the scale needs to range from the maximum of a concept to the minimum of that same concept. For example, it may be confusing if the scale moves from "extremely pleasant" as its maximum to "extremely sad" as its minimum. There are two different concepts being measured here and the scale will likely create confusion among participants. If "extremely pleasant" is the maximum, "extremely unpleasant" should be the minimum.

Depending on the purpose of the question, you may consider having an odd or even number of responses. If you are looking to *force* respondents to take a stand, you may consider having an even number of responses. Otherwise, an odd number of responses will allow for a neutral point, which can work very well depending on the type of question. Sometimes we may want to capture how participants feel or perceive an issue and whether they would lean toward supporting or rejecting it if they had to choose. In these cases, an even number of answers may be the best option for us. At other times, we may be interested in the percentage of participants who are uncertain or neutral about their position, so our best option would be an odd number of answers. It is also helpful to remember that too many answer choices (over seven options) may create confusion among participants. It is better to have five to seven options on the scale where respondents pick how they feel or think about something.

Finally, the answers should be inclusive of all options and gradually progress from one to another—as the word *scale* suggests. For example, it may not be a great idea to include only positive opinions on a topic because that excludes participants who may not feel or think positively about it. Similarly, it may be a little messy if answers jump from strongly positive to strongly negative to somewhat positive to neutral to somewhat negative. Keeping the answers gradual from strongly positive to somewhat positive to somewhat negative to strongly negative will result in a cleaner scale that allows participants to pick the answer that best represents their feelings or attitudes on some topic.

Step 9. Conduct a pretest to evaluate the instrument.

Running a pretest with a group of people is helpful in many ways. First, it will give you a sense of how much time respondents need to complete it. If your questionnaire takes an hour to complete, this may hinder recruitment. Confirming that your questionnaire does not take more than 10 to 15 minutes will increase the ease of recruiting. Second, by running a pretest, you can determine any issues in understanding questions, their sequence, or whether any questions might have caused discomfort. All of this information will help the researcher correct any glitches. Third, the participants in the pretest may volunteer suggestions about certain terms or language that will best convey accurate information. Finally, you may consider entering these pretest data in an Excel spreadsheet or other software, such as SAS, SPSS, or STATA, to further test the question construction. Sometimes, the problems that arise during data entry could have been recognized and corrected during the data collection phase.

METHODS OF DATA COLLECTION

This section introduces three major methods of data collection: personally collecting the information, computer-assisted telephone interviews, and virtual collection of data. The method of data collection is focused on the process of transforming the raw information from our participants into measurable units that can later be analyzed using various statistical tools. We are actively involved in data collection and follow specific rules on best practices to systematically and professionally collect the information we need.

RESEARCH IN ACTION 7.1

THE USE OF PROTOCOL TO ENSURE HIGH VALIDITY

In this article, we can look at the importance of the quality of data collection. The authors go into great detail about how to ensure high quality and high validity of data collection. Here, we can also note the use of a protocol with specific guidelines and training of people who are conducting the collection of information.

Source: Yamanaka, Fialkowski, Wilkens, Li, Ettiene, Fleming, Power, Deenik, Coleman, Guerrero, Novotny (2016). Quality assurance of data collection in the multi-site community randomized trial and prevalence survey of the children's healthy living program. BMC Research Notes, 1–8. DOI: 10.1186/s13104-016-2212-2

Data quality is a key priority when planning a study to guarantee appropriate results and conclusions. Detection and remediation of errors in the data collection process, whether they are made intentionally or not, promotes data integrity. . . . Quality assurance (QA) is one approach to ensure the validity of study results and preserve data integrity during the data collection process.

At the very beginning this paper is emphasizing the importance of data quality during collection. We can see how the quality of data influences validity of the study and even its integrity.

The first step to QA is developing a well-written, comprehensive, and detailed procedure manual for data collection. Poorly written manuals increase the chances of errors and risk the validity of the study. Second is developing a rigorous and detailed recruitment and training plan to enforce the value of collecting accurate data. The final step is to monitor and evaluate the process in the field and identify areas of improvement to strengthen the study's protocol.

Further on, the authors are delineating a three-step process that can ensure data quality during data collection. Initially, we start by a comprehensive procedural manual for data collection. This is followed by a rigorous recruitment and training plan for people who will collect the data. Finally, we see how active the process is when the researcher needs to monitor and evaluate the entire process in the field and constantly update and strengthen the protocols.

© iStockphoto.com/mediaphotos

So what is the best approach to data collection?

Personally Collecting Questionnaires

Personally collecting information can be laborious due to a large sample size; however, this technique is widely employed by researchers. Data collection can take place in various settings, such as schools, workplaces, malls, and other venues where people gather. Surveys might also be conducted by visiting people's homes. The researcher or researcher assistants may either approach people individually and verbally ask them the questions or allow them to read and answer on their own. Or they might choose to distribute written questionnaires to volunteer participants, allow time to complete, and then collect them.

In-person data collection is preferred because of the following advantages:

- The researcher gains the ability to note body language, record impressions, and keep records of additional information that would not be included in the questionnaire.
- It is more likely that a participant will fully complete the questionnaire, and a complete questionnaire is more valuable than a participant's partial effort.
- A longer questionnaire is easier to complete face-to-face, since participants are probably more likely to talk at length to another person than focus on a long, written questionnaire.

Some disadvantages of this method are related to cost and time limitations. To reach the desired number of participants will take time. Because it is a time-consuming process, researchers often hire assistants to aid them. Although having additional help is very useful, it will depend on the resources available.

Computer-Assisted Telephone Interviewing (CATI)

Computer-assisted telephone interviewing has radically changed the way we conduct research. CATI is used to find the right respondents or possibly to conduct the entire interview electronically. In the first approach, the computer automatically calls random numbers and invites people to participate. Once a person agrees, that telephone call is forwarded to the interviewer, who continues the conversation. One common and

RESEARCH WORKSHOP 7.2

COLLECTING DATA

Collecting your data face-to-face while recording participants' body language and other emotions and behaviors can be a gratifying process. It is always a good idea to first do a dry run with a friend who can provide feedback to improve your interviews. One of the most comprehensive guides on interviewing is the Kinsey Interview Kit from the Kinsey Institute of Research on Sex, Gender, and Reproduction (available at http://www.kinseyinstitute.org/pdf/Kinsey_Interview_Kit.pdf). The tips and techniques provided are aimed at researchers who are interviewing participants on a sensitive topic and are useful in terms of how to be unbiased, command your own body language, and collect accurate data.

The interview process is described as having three stages: the introduction or the prologue, the data-gathering phase, and the conclusion. During the introductory stage, the participant needs to be made aware of what the study is about, the aim of the research, objectives, and such. While a researcher may not be inclined to reveal the hypothesis of the study, he or she can still provide sufficient information to put the participant at ease. This is also an opportunity to reinforce the anonymity of the responses and the measures that will be taken to protect the participants' identities.

Once the participant is reassured and has gained trust, the data collection begins. At the conclusion of the interview, the interviewer may once again ensure the protection of the participant's identity and let him or her know when the results of the study will be available, if the information will be released or published.

widespread example of CATI is the customer service system of any business that channels our phone calls to the right customer service representative. In this case, the computer initiates the call.

The second type of CATI are interviews conducted entirely by computer. The participants may communicate in code by pressing numbers that are translated to words or Likert scales. All the answers are recorded and available as raw data.

Regardless of whether interviews are conducted on the phone, virtually, or face-to-face, any good interviewer has to prepare. Training sessions for research assistants are typically mandatory. Before collecting the information, the interviewer must become familiar with the content of the study, must learn what will be said to the participants, and must learn what behaviors may be interpreted as biased or judgmental. Sometimes in sensitive interviews, participants may dodge questions if they feel they are being judged. By setting a comfortable environment and tone, providing appropriate context, and avoiding any sense of judgment, this may be minimized. The interviewer is trained to be pleasant, but not lead the interviewee with verbal or nonverbal cues. Neutral language should be used in an interview to avoid hinting at preferred answers. Sometimes even a nod or a smile at an inopportune instance may be misinterpreted as a leading sign.

The protocol is similar to a checklist to organize the process. Information about the interviewee is recorded, along with the date, time, and place, as well as the length of the interview. Sometimes the interviewer may note other details. The protocol organizes the process. The interviewer typically begins with a confidentiality notification. The confidentiality statement includes background information on the study, how the data will be handled, measures to protect confidentiality, and disclosure of any potential harm that may result. When the interviews are conducted over the phone, this statement is read to the respondent; for face-to-face interviews, it is commonly provided in writing. The interview protocol is another tool that needs to be prepared in advance.

Virtual Data Collection

Online surveys are extremely popular and can vary from opportunities to provide feedback via links sent in emails to mandatory surveys in the workplace. Among the great advantages are the vast numbers of participants who may be recruited in a matter of minutes, with a much higher response rate than collecting data face-to-face or via telephone.

On the other hand, online surveys have their drawbacks. A substantial portion of the population still does not have access to the internet. This holds true within and outside of the United States. Second, participants who feel strongly about an issue are more likely to participate than those who fall in the middle. It is common to capture more positive and negative extremes, while missing the responses tending toward the middle. Third, the researcher cannot guarantee the identity of the respondent in an online setting.

RESEARCH WORKSHOP 7.3

RUNNING YOUR SURVEY ONLINE

There are several websites, such as www.surveymonkey.com, www.surveypro.com, and www.zoomerang.com, where you can easily run your own survey and collect data. Although these sites help to reach many participants and can have a high response rate, the design of your survey will impact your success.

1. Any good survey should start with a clear introduction that includes the focus of the study and reassurance of confidentiality.
2. The questions need to be consistent and visually appealing. Visual distractions or inconsistencies may result in partial completion or even refusal to participate.
3. Requiring an answer for each question before a participant can move on can also create problems. It is always helpful to add a "skip" box. This goes hand in hand with a participant's ethical right to avoid answering a question.

Summary

An experimental study works well to test the effects of some form of intervention, training, or any other idea in a group of people. Experimental studies can be conducted with a small group of people as long as the three characteristics of experimental design are used: (1) randomly assign participants in a control and an experimental group, (2) conduct a pretest and a posttest that can measure the difference caused by the experiment, and (3) make sure there are no external circumstances that can cause the differences we see in the posttest. In addition, we need to make sure we obey by the ethical rules and regulations in conducting experimental studies, such as protecting participants from harm, protecting their anonymity, and informing them of their rights, including the right to drop out of the study at any time.

This chapter introduced surveys and questionnaires. Survey refers to the method of data collection, whereas a questionnaire is the instrument containing the questions. We must determine whether the survey questionnaire for data collection has been tested for reliability and validity. If the questionnaire was never tested and we know little of whether it is a reliable and valid instrument for data collection, we refer to it as a nonstandardized survey questionnaire. Alternatively, if we are using a questionnaire that has passed the test of being a reliable and valid form of data collection, we refer to it as a standardized survey questionnaire.

When designing questionnaire questions, we start by creating a list of the most important constructs, including the list of constructs we may want to control for. Once we are certain about what we are measuring, we move on to determine how each construct may be captured in the form of a question. Each concept will require brainstorming, experiments, and even little trials to see that our questions are truly measuring what we want them to measure. We also need to consider all the possible answers to a question. Reviewing studies and literature that measure similar constructs may be very helpful. In addition, we need to make sure that our questions do not include biases, do not have meanings that can easily mislead participants, do not ask for socially desirable answers, are not double-barreled, and are not overly technical. Sometimes we may ask about a topic that is socially desirable for participants, which may cause participants to feel that they have to answer in a certain way, even if their answer is not accurate. Double-barreled questions are those questions that ask for two different terms or concepts, but include only one answer.

Our questions need to be clear and brief without double meanings or anything that can be misinterpreted by participants. Paying attention to contingency questions (questions followed up by another question) is also important because it avoids any type of distraction for participants. With contingency questions, we must also provide an option for people whose answer does not require a follow-up. Finally, it is a good idea to test the questionnaire with a group of people willing to provide feedback. This way, it will be easier to fix small mistakes and prepare the best instrument for your study.

Creating a scale to measure participant answers is an effective way to look at their attitudes, beliefs, and opinions. Scales come in different shapes and forms. Paying attention to the type of scale that best measures the constructs in our study is primary. Oftentimes, long scales may confuse participants, so we should keep them as brief as possible without interfering with what we are measuring. There are certain cases when we are interested in a neutral position from participants and allow for odd-numbered scales. Just the same, there are times when we truly want to measure how participants will rate themselves when a neutral point is not available. In such cases, we use even-numbered scales.

Researchers collect data in various forms, such as personally collecting all the information, virtual data collection, computer-assisted telephone interviews, and others. The way data are collected and the tool that will yield the best possible results should always be chosen in order to benefit the study.

Key Terms

Contingency question: follow-up questions.

Double-barreled questions: questions that combine two or more questions into one and have only one possible answer option. Double-barreled questions should be avoided because they can be quite confusing for participants.

Experimental research: conducting an experiment or intervention on a randomly assigned group in a study, while controlling findings with a control group.

Non-standardized survey questionnaire: a questionnaire that has not been tested and has not been confirmed to be a reliable and valid instrument.

Questionnaire: an instrument containing questions used for data collection.

Standardized survey questionnaire: a questionnaire that has been tested and is a reliable and valid form of data collection.

Survey: a method of data collection.

Taking a Step Further

1. What is the relationship between a survey and a questionnaire?
2. Can you provide an example of a double-barreled question?
3. What are some benefits of virtual data collection?
4. What are the steps for designing a questionnaire?
5. What are the basic requirements of experimental research?
6. What is the purpose of contingency questions in our research design? Can you illustrate with an example?

8

SECONDARY DATA

CHAPTER OUTLINE

WHAT WILL YOU LEARN TO DO?

1. Describe the benefits of using secondary data
2. Identify the major sources of secondary data
3. Explain the drawbacks of using secondary data

BENEFITS OF USING SECONDARY DATA

Secondary data refers to data or raw information collected by other researchers. All the data collection methods discussed in Chapter 7 refer to **primary data**, or data that are collected for a specific purpose by the researcher. Although highly desirable, collecting your own data may not be feasible or desirable, especially when you find yourself surrounded by raw information already collected by other scientists or institutions. Using available data implies reducing the time and costs of collecting data, so this method is quite popular. This is especially beneficial when the existing data are rich in detail and include the information the researcher is searching. For example, researchers use everything from U.S. Census data to smaller-scale institutional data to make conclusions about demographic or housing trends or various characteristics of populations, such as race, ethnicity, religiosity, or social mobility.

The following are six benefits to secondary data:

1. Availability of information
2. Opportunities for replication
3. Protection of participants
4. Time effectiveness
5. Cost effectiveness
6. Large datasets

Let us take a closer look at each of the specific benefits of analyzing secondary data.

Availability of Information

Researchers may be interested in data or information that has since become impossible to collect. The famous French sociologist Emile Durkheim used secondary data to examine government statistics on suicide from different countries. This information would have never been directly available to him, but, by accessing the secondary sources of data, he was able to see that suicide rates were substantially higher in Protestant countries. This information led him to articulate the importance of social circumstances rather than biological or individual characteristics that can cause a person to take his or her own life. It was a great discovery at the time because people believed that suicide was more of a personal disturbance of the brain, so it brought new evidence on the impact of social circumstances on the individual.

Opportunities for Replication

Replication of a study can happen in two ways: (1) we can collect primary data that are exactly like the original study, but in a different population or in a different location to see whether the findings hold true, or (2) we can use the secondary data by simply collecting additional information that may be missing from the original study.

In the first case, researchers may be intrigued by findings of an original study (either ours or another researcher's), and in order to further advance knowledge on the topic, they may want to replicate the study with another population group or at another location after some time has passed. Research in Action 8.1 provides an illustration of a replication study on personality traits and academic success conducted years after the original study. The advantage of replication studies is that the study design is completed, including research questions. Although the researcher may still need to collect additional data, the new findings will be an extension of the original research.

In the second case, we draw on the original study to find out additional information from the population. In these types of replication studies, the original researcher's findings are taken at face value and we are simply providing additional information. This information can be collected from the same participants of the original study when they are known and willing to participate or from the same type of population. For example, imagine that a researcher initially collects data on college students' likelihood to engage in academic dishonesty (e.g., cheating on an exam). This researcher may be interested in the relationship between the likelihood of engaging in academic dishonesty and high school grade point average (GPA). Once the data are collected, the researcher may not find an association between high school GPA and academic dishonesty, but is still very interested in the topic. Drawing on the original data, the researcher may go back to the same participants or another group of students from the same population and probe for more information. Some examples of additional information could be parents' marital status, parents' education level, belief in authority, level of self-efficacy, and others.

Protection of Participants

Secondary data become useful with sensitive populations of interest. To illustrate this point, we can consider victims of childhood abuse. Researchers interested in helping this population can potentially harm their participants just by interviewing them. Recalling difficult early events can cause emotional and psychological harm to the victims. Therefore, the use of secondary data, such as police reports, medical examinations, or other information gathered in the case, not only is beneficial for the intended research, but also protects the participants.

RESEARCH IN ACTION 8.1

ILLUSTRATION OF A REPLICATION OF A PREVIOUS STUDY

Source: Paunonen, S. and Ashton, M. (2013).On the prediction of academic performance on personality traits: A replication study. *Journal of Research in Personality, 47*(6), pp. 778–781.

In this article we describe a study in which we used measures of personality traits to predict students' academic performance in an undergraduate university course. Our purpose for undertaking this study was twofold. First, we wanted to determine whether we could replicate the personality-performance relations we found in a parallel study published some dozen years earlier (Paunonen & Ashton, 2001a). Second, we sought to confirm a general conclusion we made in that earlier study, and elsewhere, concerning the utility of the Big Five personality variables as predictors of important behavior outcomes. We elaborate on both of these goals in the following two sections, respectively.

> Notice how the goals of the study are presented. First, the researchers are replicating an earlier study on personality-performance relation they conducted years ago. They are drawing on their first findings to see whether the Big Five personality variables are predictors of academic performance among college students.

One aim of the present study is to provide a direct replication of some results from our earlier study of personality and academic achievement (Paunonen & Ashton, 2001a). In that study, we had access to the course grades of a large sample (N = 717) of undergraduate university students who completed a personality psychology course. . . .

In the present study, we had access to another large sample of students (N = 652) who have since completed the same undergraduate psychology course as those students in the Paunonen and Ashton (2001a) study. We were able to evaluate the replicability of the former results directly, because the criterion is student performance in the same course, taught by the same professor, based on very similar examinations, with the personality predictors based on the exact same set of personality questionnaire items (i.e., the PRF, personality research form) as in the original study.

> Both designs are the same; the samplings sizes are similar ($N = 717$ and $N = 652$). Almost everything is the same as the original study with the exception of the student sample. The professor and the measurements of personality traits are the same and the exams are similar.

Time Effectiveness

It goes without saying that collecting primary data requires a great deal of time. Collecting quantitative or qualitative information can take a few months to a few years; the process is very time-consuming. This is especially true for larger studies. In fact, very large studies require a team of people to collect the data. It is almost impossible for one person to collect that much data in a timely manner. To analyze already-collected data, however, is much easier and does not require the effort of an entire team, although using a team is preferred when possible. Using secondary data eliminates the time and effort of data collection and the researcher is able to move directly to cleaning, coding, and analyzing the available data.

Cost Effectiveness

Data collection is not simply time-consuming. It also requires a large budget that may not always be available to graduate students or to other researchers who are just starting out. To collect even a small amount of data, a researcher needs to have funds for proper technical tools and to pay additional people for collecting information, to name a few. When using secondary data, these means are not necessary to complete the project, making the study more cost-efficient.

Large Datasets

Another benefit to using secondary data is the ability to use a large amount of data that would not have been otherwise possible. Larger datasets like census data allow a researcher to analyze huge amounts of data, making it easier to generalize his or her findings to the general population. Finding sources of data already collected from either an institution or an individual is a great opportunity to test new theories on large amount of data. In addition, with the advancement of technology, researchers are now able to look at data from internet sources that were previously unavailable.

Data scraping, or extracting large amounts of information from websites into a readable spreadsheet like Excel, from social media sites, business sites, or any other websites of interest gives the researcher an opportunity to access information in a way that was not even imaginable a few years ago. Data scraping allows us to research such things as consumer purchasing behaviors, political views, and friendship connections. The technique is simple to learn and do on your own: You download the information from the sites you are interested in and collect the data You need. However, there are also companies that can do the scraping for you. After the data are collected, you can begin your analysis.

RESEARCH WORKSHOP 8.1

SCRAPING DATA

Data web scraping is a magnificent tool that allows you to collect information from almost any source—from social media websites, such as Facebook and Twitter, to businesses like Amazon .com—depending on what you are interested in pursuing. Learning how to scrape data is easier than ever and there are many useful sources that will help you accomplish this.

- The following is a free course on scraping and data mining from Udemy: https://www.udemy .com/scraping-and-data-mining-for-beginners-and-pros/learn/v4/content
- The following is a free guide on data scraping, with software: http://www.datascraping.co/doc/
- Data Scraping Studio helps collect information from various websites. Information on this service can be found at http://www.datascraping.co/. In addition, it may be beneficial to ask a graduate student or a professor in your Computer Science program at your university for help. They may be able to direct you to the right sources or even help you with data scraping.

Finally, the following is a list of some books on data scraping for beginners:

- *Web Scraping With Python: Collecting Data From the Modern Web* by Ryan Mitchell
- *Automated Data Collection With R: A Practical Guide to Web Scraping and Text Mining* by Simon Munzert, Christian Rubba, Peter Meibner, and Dominic Nyhuis
- *Web Scraping With Excel* by David M. W. Phillips
- *Website Scraping and Tools for Beginners* by Leif Izland
- *Social Media Mining With R* by Nathan Danneman and Richard Heimann

MAJOR SOURCES OF SECONDARY DATA

There are a number of available sources of secondary data that are collected using strict research methodologies, such as data collected by government agencies or academic institutions. Other sources of secondary data may not strictly follow methodological rules and regulations, but are available for learning purposes. Although we have an abundant amount of data at our fingertips, it is important to carefully examine these secondary sources by initially looking at their methodologies. How were the data collected? What were the characteristics of the population frame, and how was the sampling frame determined? What type of sampling was used? How many respondents participated? We also need some information on data availability, such as the possibility of using the data for our own analyses and whether there are applicable terms and conditions. This information is quite useful in determining what source we will be using and whether the available data will be sufficient for testing our research hypotheses or questions. Let us briefly look at some common sources of secondary data available.

Government Statistics

All governments collect various forms of information about their citizens. We are familiar with census data in the United States, but every other country has its own similar archives of information about its citizens, collected over time. This large database is available to researchers in its raw form and includes basic information, such as birth rates, death rates, marital status, number of children, annual income, crime rates, and employment status. By investigating datasets on specific populations, researchers are able to examine almost the entire population rather than just a limited sample. Often scientists are interested in demographic trends over time, such as the number of children per family, unemployment rates, or crime rates in a specific area. Government statistics are some of the most comprehensive and oldest data available for most populations.

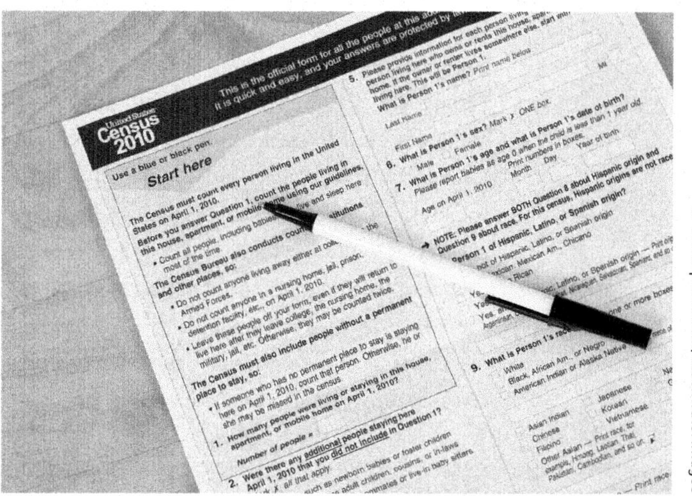

The Census Bureau collects important data on various aspects of life that can be used as secondary data.

Similarly, other governmental agencies have created large archives of information, collected over years. Datasets such as these are available from the Centers for Disease Control and Prevention, the National Institutes of Health, the Department of Housing and Urban Development, and the Department of Education. These databases go beyond basic census data and offer more details about specific topics of interest. The information is most often specific to the purpose of the agency, such as health, housing, and education.

Research University Data

Scientists can access available datasets through research universities. Research universities often create institutes dedicated to data collection in a specific field in the form of valuable longitudinal data made available to other researchers. One of the largest longitudinal studies of this kind is the Panel Study of Income Dynamics from the University of Michigan, which is freely available to the public. This significant project has followed 18,000 individuals from 5,000 families in the United States since 1968. The database includes the topics of demographics, health issues, child development, employment, and marital status. This type of resource is important to scientists since it includes information that due to cost and time restrictions may not otherwise be available.

Similarly, the German Socio-Economic Panel includes information collected from 1984 to 2007 on 20,000 individuals from 12,000 families. The data include household composition, employment, health, and other measures. The Swedish National Data Service also includes large datasets from the 1980s on various aspects of household characteristics. More recently, other organizations have followed in the footsteps of the University of Michigan with their own longitudinal studies, including the Household Income and Labor Dynamics in Australia, begun in 2001, and the Survey on Household Income and Wealth in Italy in 2008. Other research universities provide large datasets that are available to the public, in which studies based on these data become property of the university and are available to anyone who is interested in reading them.

Another possible option is the data collected by a specific researcher at one of these research universities. Oftentimes we become fascinated with a study and are truly interested in getting the data that the researcher has collected. The researcher in charge of the data collection is often the contact person for getting the needed information on the methodology of data collection as well as the actual dataset. The contact information of such researchers is commonly provided in the articles or books they have published. By contacting them, we may get the necessary information about the rules and regulations on how to access their dataset—if that is a possible option. Often researchers have a personal research interest in having

Universities are the most common research institutions. Dedicated, individual researchers collect rich information on different topics.

other researchers analyze their data and are eager to provide all the information needed, but this remains their prerogative.

Institutional Data

Universities, like other sizable institutions, have additional data besides those collected by their own scientists, such as information on students, faculty, and other employees. Any large organization, such as private and public schools, universities, hospitals, social media, large business, and others, also collect information about their clients and employees. Although their data are often in raw form and were not collected for research purposes, they are a potential source of data for scientists interested in studying topics that relate to the type of data they have available. Some topics that could benefit from this type of data are work satisfaction, salary classification by position or rank, the population that the organization serves, consumers who purchase goods from a particular business, and many others. Institutions collect information not simply about their clients and employees, but also about these individuals' families, past histories, and other characteristics specific to that institution. For example, an organization that provides services to the homeless population will collect data over time about them. Such data could include gender, income, education, family history, history of drug abuse, mental health, marital status, and many others. A supermarket with a membership system—which is true of most supermarkets—will have information on consumers' age, income, gender, marital status, parental status, as well as their purchasing habits. As you can tell such information is quite rich in detail and offers great opportunities for researchers. Someone interested in patterns of homelessness could really benefit from the information collected by a homeless organization. Alternatively, a researcher interested in nutrition or even someone conducting market research for the food industry could use the data provided by a supermarket.

Online sources have amazing potential to become the leading source for getting secondary data.

Online Sources

Popular social media hubs, such as Facebook and Twitter, where people congregate, review, and comment, are virtual gems for researchers. Some of the information may be demographic, such as gender, sexual orientation, age, political orientation, or other information. Other data take the form of narratives pertaining to specific issues. Technology has changed the way we behave, think, and act and has brought drastic change to the way we conduct research. Never before has such rich, qualitative information been made so readily available for research purposes.

Even by exploring reviews on a product, procedure, event, or experience, we can easily collect abundant information. For example, researchers Shannon Hughes and David Cohen (2011) were interested in the uses and side effects of psychiatric medications and started investigating broad-based information on two different drugs—one antipsychotic and one antidepressant. They analyzed 5 years of information about these two drugs based on consumers' online reviews. They discovered that people used these psychiatric medications for off-label uses to treat migraines, insomnia, and various developmental disorders rather than their intended use. This study would have been more difficult to complete without the use of secondary data.

With its easy availability, online information provides a wealth of secondary data to bolster new and valuable research. Table 8.1 shows a list of helpful resources that researchers can use to collect secondary data.

TABLE 8.1 ■ Resources for Secondary Data
www.ahrq.gov The Agency for Healthcare Research and Quality
www.kaiserpermanente.org Large Health Maintenance Organizations
www.cochrane.org/index.htm The Cochrane Library
www.cdc.gov/nchs The U.S. National Center for Health and Statistics
www.who.int/en/ The World Health Organization
www.mayoclinic.com The Mayo Clinic
www.ucsf.edu University of California, San Francisco School of Medicine
www.census.gov The U.S. Census Bureau
www.hhs.gov The U.S. Department of Health and Human Services
www.kaggle.com Kaggle
www.drivendata.org DrivenData
https://archive.ics.uci.edu/ml/datasets.html University of California, Irvine, data repository
http://www.pewresearch.org/data/download-datasets/ Pew Research Center
http://enigma.io/publicdata/ Enigma

Source: Using an existing data set to answer new research questions: A methodological review, by Daniel M. Doolan and Erika S. Froelicher *Research and Theory for Nursing Practice: An International Journal*, 23(3), 2009.

RESEARCH IN ACTION 8.2
AN ILLUSTRATION OF AVAILABLE DATASETS

Keeping it fresh: Predict Restaurant Inspections—Competition from Yelp in Driven Data

Sources: Peter Bull, Isaac Slavitt, and Greg Lipstein, "Keeping it Fresh: Predict Restaurant Inspections," April 2016.

Peter Bull, Isaac Slavitt, Greg Lipstein. "Harnessing the Power of the Crowd to Increase Capacity for Data Science in the Social Sector," arXiv:1606.07781 [cs, stat], Jun. 2016

The City of Boston regularly inspects every restaurant to monitor and improve food safety and public health. As in most cities, health inspections are generally random, which can increase time spent on spot checks at clean restaurants that have been following the rules closely—and missed opportunities to improve health and hygiene at places with more pressing food safety issues.

This is an example of a description of data in one of the previous competitions in Driven Data. Initially we are introduced to the regular procedure of how health inspections are done in the city of Boston.

Each year, millions of people cycle through and post Yelp reviews about their experiences at these same restaurants. The information in these reviews has the potential to improve the City's inspection efforts, and could transform the way inspections are targeted.

A team of Harvard economists and Yelp—with support from the City of Boston—are co-sponsoring this competition to explore ways to use Yelp review data to improve the inspections process. We are looking for your help to achieve this goal.

We are also introduced to the fact that many customers also post Yelp reviews about the same restaurants where inspections are done. If the reviews were used properly, the City of Boston could have an easier time targeting their next restaurant for health inspection.

(Continued)

(Continued)

Winning algorithms will be awarded financial prizes—but the real prize is the opportunity to help the City of Boston, which is committed to examining ways to integrate the winning algorithm into its day-to-day inspection operations.

The goal for this competition is to use data from social media to narrow the search for health code violations in Boston. Competitors will have access to historical hygiene violation records from the City of Boston—a leader in open government data—and Yelp's consumer reviews. The challenge: Figure out the words, phrases, ratings, and patterns that predict violations, to help public health inspectors better perform their jobs.

The introduction also tells us that there is a team of Harvard economists who are attempting to explore these Yelp reviews to improve the inspection.

This is a paid competition, so the description mentions that.

The goal of the competition is made clear in this next paragraph where we see that an exploration of words, phrases, rating, and patterns—most likely a combination of all of them—are the ingredients to create the right algorithm that can help public health inspectors.

The competition opens Monday, April 27th and will accept submissions for eight weeks. Submissions will be evaluated on fresh hygiene inspection results during the six weeks following the competition; after that, the prizes will be awarded. Your submission will not only put you in the running for the prize—it has the chance to transform how city governments ensure public health.

It is clear here that Yelp has made their reviews available to anyone for further exploration. Similar procedures are done by various companies that are making their information available in the hopes of creating a better future for all of us. This is where secondary data can truly be useful.

Some other increasingly popular data sources are websites such as Driven Data, Kaggle, and Enigma. The first two companies host data science competitions for novices, enthusiasts, and experts of data science. In other words, if a company or a researcher decides to make their datasets available to public for scientific analyses, they release their data to Kaggle or Driven Data. Some businesses can offer prizes, such as monetary compensation for competition winners, while others simply make their data available for learning purposes. Each available dataset has a lengthy explanation of the methodology on how data were collected and the procedures followed. This information helps determine the quality of the dataset, the methodology used, and the methodological standards followed.

DISADVANTAGES OF SECONDARY DATA

Although we are aware of the benefits of saving time and costs in data collection, there are a few disadvantages of using secondary data as well. They include the following:

1. Uncertainty of constructs
2. Ambiguity of measurement error
3. Passage of time

Let us explore each of these disadvantages in detail.

Uncertainty of Constructs

As you may recall from previous chapters, researchers create constructs and express them in the form of variables. Although most researchers rely on previous studies to measure these constructs, there is always room to improve these instruments. New studies and findings continually improve previous construct measurements, so using secondary data can limit a study to the results of an older or possibly outdated data collection instrument. Because the original study had a different focus, some constructs may have been of a higher importance than others. When using secondary data, you must consider whether the original construct fits the focus of your current study, as uncertainty about the measurement of the original construct can cause flaws in your study's results. In a hypothetical scenario, imagine how you are very interested in understanding what type of crafting activity is most popular among graduate students. More specifically, you are thinking that knitting is perhaps the most common crafting activity of female graduate students. Instead of collecting your own data, you may be thrilled to hear that someone else already has a lot of information on crafting activities of graduate students. This other researcher, however, was focused on understanding the type of crafting stores graduate students frequented the most. When you look at the questions, you notice that there is a question about the type of crafting these students engage in, but the question divides crafts and artistic activities into five categories: textile crafting, paper crafting, fashion crafting, decorative crafting, and visual crafting. The survey includes a lot of detailed information about the stores frequented, but it does not have much more on the crafting types. This can be a problem for you if you were considering using these secondary data because the division between the artistic and crafting activities is too large and inclusive of many different types of crafts. Therefore it will be very difficult to even test your hypothesis about knitting. Knitting will be included in the same category as sewing and quilting, none of which were part of your hypothesis. This secondary data cannot quite be used for your study because of the differences in construct measurement. This also relates to the next disadvantage of secondary data: ambiguity of measurement error.

Ambiguity of Measurement Error

When we are highly involved in the process of data collection, we are aware of every aspect of the process, including what might have been measured differently, how we might have improved our procedures, which parts of the study are strongest in validity and reliability, and topics that are best suited for further research. This type of knowledge is crucial to data analysis. The researcher using secondary data cannot benefit from this inside knowledge and may therefore have difficulty with evaluation. In fact, when we use secondary data, we often take the information at face validity because we lack a good sense of measurement error. Let us assume that we are interested on where crafters sell their finished products. We are aware that people who paint or quilt or make jewelry often sell their products, but where? What if we were to include a question about where people sold their finished crafting projects and included a number of options, such as eBay, Amazon, Etsy, personal blogs, and personal websites. We run the survey and collect information from 200 participants and as the data are coming in, we see that many people who sell their products selected none of the options we put together as selling venues and wrote about community fairs, church fundraisers, community events, and charity events in the comment box. Clearly, there was a problem with this measurement in this case. So if another researcher were to use the data from this same survey, they could end up with measurement error.

Passage of Time

Even when the secondary data source is accurately measured and it truly fits our own research, there is a possibility it is outdated. Depending on the study's focus, some information becomes old quite quickly, whereas other information may remain relevant. Our daily lives change with tremendous speed and researchers do their best to keep up with it. For example, census data are collected every 10 years. While it makes sense to use that information 1 to 3 years after the data are collected, the same information gradually becomes outdated and less accurate.

However, the passage of time can also work to our benefit if we are interested in longitudinal studies and want to see how a concept has changed over time. For example, if we wanted to see the number of children in families over a period of 50 years, we could look at census data from 50 years ago to the present to see these trends. In a case like this, we would not be concerned about outdated data.

ETHICAL CONSIDERATION 8.1

STUDY PARTICIPANTS AND SECONDARY DATA

When it comes to secondary data, we may feel as if ethical concerns about participants are nonexistent because we do not have direct contact with them. However, there are still some major issues concerning ethics when it comes to using secondary data. Participants are completely anonymous to researchers in such cases, which makes us feel that we do not need to worry about any violations. That said, we should still be careful in reviewing the ethical decisions of the original study and make sure that the primary study conformed with all ethical regulations. In addition, we need to carefully review how much of the information from this data we can reveal in our study as well as the confidentiality level. Sometimes, even though we are not directly involved with participants, we may still harm their confidentiality. If the secondary data we are accessing include confidential information about participants, or information that can identify the respondents who took part in the study, we are still responsible for protecting their identities and anonymity even though we were not the primary data collectors. Commonly, this information should be included from the source of primary data, but it is important to be aware of this possibility from the very beginning.

In addition, most research institutes, universities, government, and nonprofit organizations have specific rules regarding how the data are distributed for further research. There are cases when the data are freely available for public use and there are other scenarios when a fee is applicable to gain access to a dataset. This is straightforward for the established sources of secondary data, but the situation becomes more vague when scraping data from websites. For example, taking information from a specific website, analyzing it, and publishing our analysis in the form of an article, book, or conference presentation may violate copyright laws from the website from which we took the data. In such cases, it would be best to ask for permission from the website before attempting to make our work public.

Summary

Using secondary data, which have already been collected by another researcher, can be useful, practical, and beneficial depending on the study. Although many researchers use primary data, the use of secondary data has always been prevalent in research. There are six benefits to using secondary data: (1) availability of information, (2) opportunities for replication, (3) protection of participants, (4) time effectiveness, (5) cost effectiveness, and (6) large datasets.

Using information from another researcher can make our work easier, provided that the variables are measured the way we would like to measure them. Using secondary data also gives us opportunities to replicate already conducted studies, which can ensure that the primary findings are applicable in other populations or other locations, even after some time has passed. We may also add more information to the original data. Another benefit of using secondary data is time effectiveness. Collecting primary data takes a lot of time that may not be available to the researcher. When a dataset already exists on the same topic, we can save some needed time and move on to the analysis. Similarly, collecting data is costly because of the labor and tools involved. When we use secondary data, these costs are no longer a concern for the researcher. Finally, with secondary data, we are able to use large datasets, much larger than we would have been able to collect on our own. These days, we are able to scrape information from public resources and transform that information into a dataset of our choice.

Despite the obvious advantages of using secondary data, there are a number of concerns that apply as well. Some of these disadvantages are (1) uncertainty of constructs, (2) ambiguity of measurement error, and (3) passage of time. When using secondary data, we may not know how the constructs were developed and whether they were developed in a manner that fits our study. This is particularly true when we use data that were collected a long time ago. Additionally, we may run into data measurement errors. When collecting primary data, we know how the information is gathered and are likely aware of the types of measurement errors we are facing. Alternatively, when we use someone else's data, we may not have that insight and face the possibility of unknown measurement error. A final problem with using secondary data has to do with the passage of time. In some longitudinal studies, it may be beneficial to use data that were collected a long time ago. However, that may not be the case when we wish to analyze current trends.

Key Terms

Data scraping: extracting large amounts of information from websites into a readable spreadsheet like Excel.

Government statistics: statistics collected by governments all around the world. Such data include information about births, deaths, demographic trends, income, employment, and other useful information.

Institutional data: information collected by various organizations about their employees, consumers they serve, and other characteristics of the institution. Just like the government statistics, these data include demographic information that pertain to the specific institution.

Primary data: data that are collected from the researcher or the team of researchers directly.

Replication: conducting a study exactly as it was conducted the first time. Sometimes the replication can happen with minor changes, and others without any differences from the original study.

Secondary data: data that are used for analysis by the researcher or the team of researchers, but they were not involved directly in the data collection process.

Taking a Step Further

1. How do secondary data differ from primary data?
2. What are some disadvantages of using secondary data in our studies?
3. What are some types of data we can collect through data scraping? Illustrate with an example.
4. If we scrape the data we need, what makes data scraping a secondary data source and not a primary one?
5. How can we protect participants when we are using secondary data and technically have little (if any) interaction with the study's participants?
6. Why is the passage of time important to consider when using secondary data sources?

9

ENTERING AND ORGANIZING QUANTITATIVE DATA

CHAPTER OUTLINE

WHAT WILL YOU LEARN TO DO?

1. Explain the purpose of entering and organizing data logically
2. Summarize the steps needed to prepare to enter data
3. Describe how to organize and input variables and their information

THE PURPOSE OF ENTERING AND ORGANIZING QUANTITATIVE DATA

The completion of data collection is an immense source of fulfillment and gratification for any researcher. The work you put forward has paid off and you have collected and are in charge of this new information in your hands. The process of cleaning and organizing the data is similar to cleaning and organizing anything else in your life, from your dorm room or apartment to your car. There are two principles involved: (1) group all items that are classified under the same category in their designated place, and (2) remove unnecessary items, such as dust, dirt, debris, and other non-pleasantries. These same principles are applied to our collected data. We are hoping to (1) place the results in various categories where they belong and (2) remove or fix any information that has the potential to intrude into the study or our analysis. These steps are necessary for the analysis. Otherwise we would be left with a lot of information, but little understanding of its content.

Once I met someone interested in horseracing—he was fascinated by each race, the odds, and people's betting habits. With little awareness of his ultimate goal, my friend started going to the horse races with only one intention: to collect information. He would purchase the daily racing bulletin listing all the odds and horses' characteristics and would then simply document the results once the races were over. When I met him, he had collected five bins in his basement full of race results, but had no clue what to do with them. Without a way to organize the results to begin to interpret the data, the data will remain just that—bins in the basement. The very first step in organizing data is to enter them in a readable and meaningful form. But, you may ask, what happens now? How do I make sense of it all?

PREPARING FOR DATA ENTRY

The task of entering data can be repetitive and time-consuming; however, it is a crucial part of the research process. Even a small mistake in data entry can impact your outcomes. If you collect your data via face-to-face interviews, phone interviews, or other means, you need to determine a way to enter that data into a comprehensible format. If you use an online data collection system, such as SurveyMonkey, Qualtrics, or Esurvey Creator, you do not need to worry about entering your data, but can readily access a formatted data file. In this case, you would need only to continue with the process of cleaning. The following are key things to keep in mind while entering data.

Logical Formatting

© iStockphoto.com/Rawpixel Ltd

Bringing the written words into a readable spreadsheet is a key part of the data collection that demands so much attention to details.

Taking a critical look at your survey questions before beginning your data entry will save time and unwanted frustration. Determine how the variables can best be organized to ensure that the data analysis phase is simple and straightforward. For example, which variables should be entered in the first, second, and third columns of a spreadsheet? You may be inclined to follow the order in which the questions appear on the questionnaire; however, this is the time to critically assess whether the original order of the questions will, in fact, be optimal for the next phase of the process.

Sometimes, problems you encountered during data collection will alert you to any restructuring that might need to be done to ensure smoother data entry. If the question order felt awkward during data collection, it will likely be awkward in the data analysis phase as well. For instance, if you asked participants to identify their gender at the beginning of the survey, then asked about their race and ethnicity somewhere in the middle of the survey, and finally asked about their age and income toward the end of the survey, then it may have already been clear to you that this order is a little distracting and disorganized. Does it work better to group all of the demographic questions together? The logic of the original questionnaire should be assessed before data entry may begin.

Software Packages

It almost goes without saying that the data will be organized into a spreadsheet. Researchers often enter their data directly into their software of choice, because if a file needs to be converted from one program to another it may be corrupted. Common programs for entering and analyzing quantitative data are Excel, Statistical Package for Social Science (SPSS), Statistical Analysis System (SAS), statistics and data (Stata), software environment for statistical computing and graphics (R), software for statistical analysis (Minitab), statistical discovery software (JMP), and many others.

Some software programs are better for analyzing rather than entering data, but examining your options before you start is a wise decision. Although it may not be important which software you use for entering data, it is quite different when you consider

FIGURE 9.1 ■ Data Entry in R

	x	y	z	a
1	1	4	-2	5
2	4	16	1	17
3	6	24	3	25
4	8	32	5	33
5	10	40	7	41
6	14	56	11	57
7	17	68	14	69

FIGURE 9.2 ■ Data Entry in SPSS

	fathers	gender	ideajob	major
1	5	1	7	2
2	10	1	2	1
3	4	1	1	6
4	6	1	1	6
5	3	1	7	1
6	8	1	2	1
7	4	1	2	4
8	3	1	2	4
9	1	1	5	1
10	10	1	1	2
11	10	1	2	1
12	8	1	2	4
13	9	1	2	1
14	8	1	2	5
15	5	1	2	1
16	5	1	7	10

FIGURE 9.3 ■ Graphs From R-Studio

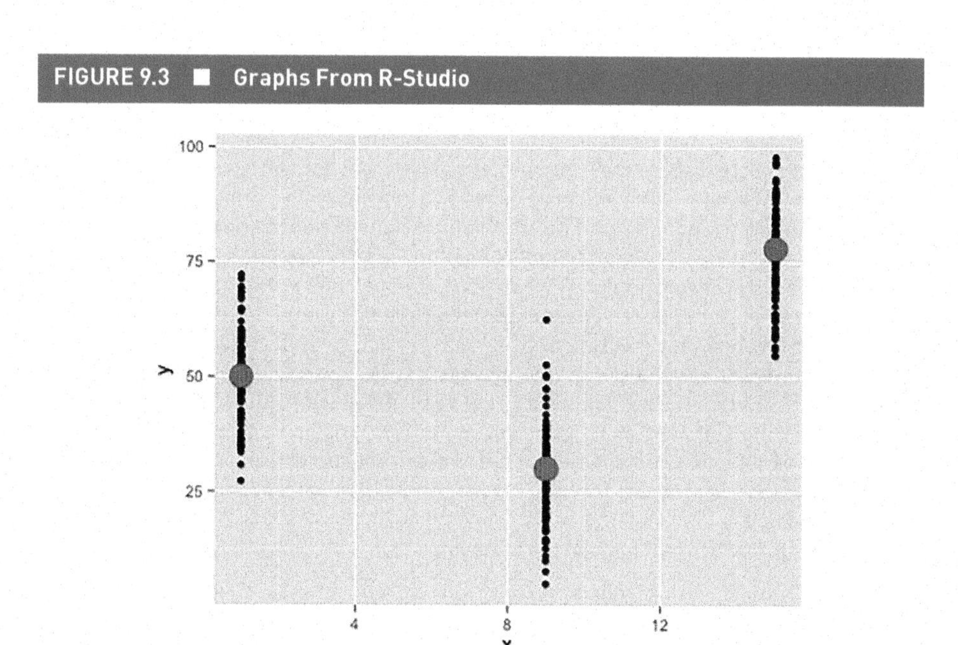

which software to use for data analysis. Therefore, it may be useful to do some internet research and determine which type of spreadsheet, like Excel, is the most compatible to be imported into the analysis software of your choosing. For example, an Excel spreadsheet saved as.csv (comma delimited) is easily imported into R and R-studio, but the import, although possible, may not always be smooth when importing into SPSS. These types of details will make the entire experience easier if you think ahead.

RESEARCH WORKSHOP 9.1

EXPLORING SOFTWARE PACKAGES

Most software packages have a trial version that lasts from a couple weeks to a month. R is the exception with a free and collaborative package that has a fabulous graphic visual package called R-studio. R-studio allows researchers to visualize their work as they code and analyze their data, so they can have a better understanding of the dataset and can simultaneously view the relationship between variables. Perhaps because it is free, there are also many available tutorial videos and websites for R and R-studio.

(Continued)

(Continued)

There are forums where people discuss the pros and cons of different software, and it is worthwhile to familiarize yourself with a few different programs to see which one you like best. Downloading the trial versions and watching the tutorials are the first steps to understanding the basic features of each package. Which one feels more comfortable for you? What is the learning curve? What types of graphs does each program produce? What types of visual interpretations do the programs provide that support your study? Which one is supported by your university and is either free or inexpensive if purchased through your institution?

Another source of information on software is your professors. They will usually have distinct reasons for selecting their software—ask them what these reasons are. It could also be helpful to know someone who uses the software and can assist you in your analysis. When it comes to learning about how to use various programs, a number of training resources ranging from short YouTube videos to entire classes are available for free. Classes may be found through Udacity, Khan Academy, Coursera, and Udemy.

Free courses that explain how to use R include the following:

- https://www.udacity.com/course/data-analysis-with-r--ud651
- https://www.udemy.com/machlearn1/
- https://www.coursera.org/course/rprog
- https://www.edx.org/course/explore-statistics-r-kix-kiexplorx-0

Free courses that explain how to use Excel in your data analysis include the following:

- https://www.coursera.org/learn/analytics-excel
- https://www.udemy.com/quick-start-excel-getting-started-with-excel/
- https://www.udemy.com/getting-started-with-excel-20132/
- https://www.udemy.com/ultimate-guide-microsoft-excel-2013/

Free courses that explain how to use SPSS include the following:

- https://www.udemy.com/spss-basics/

Please note that these free courses often change, as additional courses become available, so it is a good idea to check what courses are available every now and then. Courses in Coursera are usually available for a brief time when the course is running, and unless you download the videos, you may not have access to those courses later on. Courses in other platforms, such as Udacity and Khan Academy, are more consistent.

Unique Identification

Certain ground rules need to be established before data entry begins. A first rule of thumb is to assign a unique identification number to every participant or case in your study. Each complete questionnaire requires an identification number, which should be listed in the first column in your spreadsheet. Each column will represent a variable, but your first variable must be the identification numbers. This way, every row will represent one participant or one case. If you have more than 100 cases, you may want to consider the numbers of digits at the beginning of identification (for example, 001, 002, . . . 010).

ETHICAL CONSIDERATION 9.1

PROTECTING THE ANONYMITY OF PARTICIPANTS

Protecting the anonymity of participants is a crucial part of the data entry process. We promise participants that their identity will not be disclosed and it is because of this promise that they provide the information we are seeking. During the data entry process, we can fulfill this promise by creating either numbers or fake names to identify one participant from another. There is no circumstance that allows the use of participants' real names or publishing any of their identifying information.

Sometimes we may think that the information they have disclosed is not harmful and we cannot imagine that someone may be hurt by it. This is not the time to trust our judgment on what may or may not be harmful to someone. Even some information that may seem benign to us, like the number of drinks a person may consume in one sitting, may have severe consequences if disclosed for that participant. That is why we promise all respondents in writing that we will not distribute their personal information before they participate in our study. During the data entry process, we can keep this promise and enter the information in a way that cannot be associated with participants.

Coding and Codebooks

A number of other minor decisions need to be made before data entry can begin. First, you will need to determine how the responses to specific questions will be represented on your spreadsheet. By coding the information, we can simplify raw information in a way that is easier to analyze. In practice, this typically means translating words into numbers. For example, when we ask a participant to identify gender, they answer either (A) female or (B) male. We need to decide how to code this raw information into a numerical value on the spreadsheet. We might decide to enter 0 for participants who chose female and 1 for male. This is an example of a **nominal variable**.

Nominal variables are a subtype of **categorical variables.** Categorical variables are variables with answers that can be organized around various, often limited, categories. Nominal variables, also called *qualitative variables*, are one of the two types of categorical variables. The terms *nominal* and *qualitative* are used interchangeably to describe those variables where answers are not mathematically related to each other. For example, we cannot add one female (value = 0) and one male (value = 1) and say that one female plus one male equals 1. That expression is meaningless for a nominal or qualitative variable. Gender, race and ethnicity, religious affiliation, and sexual orientation are some examples of nominal variables.

Another type of a categorical variable is an **ordinal variable.** This refers to responses that do not have a numeric value, but instead represent an ordered sequence of responses. Let us consider the following example where the variable has four possible answers:

(A) strongly agree

(B) somewhat agree

(C) somewhat disagree

(D) strongly disagree

Again, we need to assign a number to each response (strongly agree = 4, somewhat agree = 3, somewhat disagree = 2, and strongly disagree = 1). This type of categorical variable has some degree of relationship between answers. The answers in ordinal variables may indicate a comparison between each other. So we can say that "strongly agree" is a higher degree than "somewhat agree," but we still cannot add, subtract, or mathematically compare the relationship between these answers. In ordinal variables, we may retain a numerical order because "strongly disagree" should be clearly differentiated from "strongly agree." To express the difference, we may consider coding responses from 0 to 3 or from 1 to 4, where 1 represents one extreme and 4 represents the other.

A **numerical variable** is the third type of variable that is expressed with numerical values in the original data, such as age, income, or height. Numerical variables are also known as *quantitative variables*. We may decide to initially enter these answers just as they are reported. However, that is not always the best option.

For example, say we are studying exercise habits and have collected information from a participant's treadmill activity. We measured the time of day the participant runs on the treadmill, the length of time he or she runs, and his or her average speed. We may have a time of 5:00 am, a length of time of 45 minutes, and an average speed of 2 miles/hour for one participant. These are each measured in numbers, but use different units. If we enter this information as is, we cannot interpret it unless all the data are entered in consistent units.

The entire dataset must be consistent because we will be analyzing this information. This means that we will use mathematical and statistical calculations to interpret the relationships between variables, and if units are inconsistent, we will not be able to produce any meaningful information out of the data collected. It will be as if we are comparing oranges, apples, and tangerines. Therefore, we may decide that 1 means 5:00 am and 2 indicates 6:00 am. Similarly, we may decide that a speed of 1 mile/hour is indicated by a 1 and a speed of 2 miles/hour by a 2 and so on. We may split the time within one hour into 4, so 15 minutes is assigned a 1, 30 minutes is coded as 2, 45 minutes as 3, and one hour as 4. Determining how to represent every single variable in a numeric fashion is a fundamental part of coding.

There are two types of numerical variables: **ratio** and **interval.** Interval variables cannot express a ratio between the numbers because of the lack of a *true zero*. The measurement of temperature is a classic example used to explain the difference between ratio and interval. Temperature does not have a true, absolute zero because we always have a temperature. There is never a time when temperature does not exist. Even if the temperature is 0 degrees Celsius outside, that is still 32 degrees Fahrenheit. The temperature is also a good example because we cannot state that it is twice as cold today as it was yesterday. Since there is no real ratio we can apply to the temperature, this is an example of an interval variable.

A ratio variable, on the other hand, has a true zero point. For example, if we ask participants, "How many children do you have?" they can answer "0." That is a true zero point because it is common for people to not have children. Identifying each variable, whether it is nominal, ordinal, or numerical, is necessary for running any analysis because of the various analytical requirements of the variables.

A **codebook** is the key to the codes in your study and how they correlate with participant responses. The codebook is an extremely important document to be kept in a separate and secure file. Some of the coding information will be available in your dataset, but much of it may be lost over time. Keeping the codebook secure is essential because it holds the key to your research process, your original coding decisions and your reasoning behind them, and how your code may be translated or manipulated into a new code format if it becomes necessary later on. Without the codebook, your research and the information it generated may be lost. It is advisable to include as much detail as possible about your coding process.

Missing Data

Ideally, your participants answered each question and no information was missing. As much as we prefer this scenario, it is more often the case that some people fill out half of the questionnaire or leave at least one question unanswered. Since this may be the case in your own research, you should make a few advance decisions about how to handle missing data in the data entry process.

RESEARCH IN ACTION 9.1

ILLUSTRATION OF A CODEBOOK

Source: International Social Survey Programme: Family and Changing Gender Roles IV: ISSP, 2012 by Leibniz Institute for the Social Sciences

Parts of the following codebook are taken from the Family and Changing Gender Roles International Social Survey 2012, conducted by Leibniz Institute for the Social Sciences. This is part of a longitudinal study that began in the early 1970s and is still ongoing. The scope of the study is to collect information on various social issues from different populations in Europe, Asia, and Latin America. The following are some questions taken from their codebook. Please note how every answer has a number in front of it that indicates how the information is coded in the spreadsheet. The codebook is key information for any researcher who is interested in investigating a dataset.

Q1

To begin, we have some questions about women. To what extent do you agree or disagree . . . ?

(PLEASE TICK ONE BOX ON EACH LINE)

Q1a

A working mother can establish just as warm and secure a relationship with her children as a mother who does not work.

1 Strongly agree

2 Agree

3 Neither agree nor disagree

4 Disagree

5 Strongly disagree

8 Can't choose, KR: don't know, refused

9 No answer

in Spain (ES):

0 NAP: question asked with 4-point scale

Q18

How do you and your spouse/partner organize the income that one or both of you receive? Please choose the option that comes closest.

(PLEASE TICK ONE BOX ONLY)

0 NAP, no partner (code 3 in PARTLIV)

1 I manage all and give partner his share

2 Partner manages all and gives me my share

3 We pool all money, each take out

4 We pool some money, rest separate

5 We each keep own money separate

6 TW: My son or my daughter-in-law manage the money

8 Don't know, KR: don't know, refused

9 No answer

Q19

In your household who does the following things . . . ?

(PLEASE TICK ONE BOX ON EACH LINE)

Q19a Does the laundry?

0 NAP, no partner (code 3 in PARTLIV)

1 Always me

2 Usually me

3 About equal or both together

4 Usually my spouse/partner

5 Always my spouse/partner

6 Is done by a third person

8 Can't choose, KR: don't know, refused

9 No answer

Q21

When you and your spouse/partner make decisions about choosing shared weekend activities, who has the final say?

(PLEASE TICK ONE BOX ONLY)

0 NAP, no partner (code 3 in PARTLIV)

1 Mostly me

2 Mostly my spouse/partner

3 Sometimes me/sometimes spouse, partner

(Continued)

(Continued)

4 We decide together

5 Someone else

8 Don't know, BG: can't choose, KR: don't know, refused

9 No answer

Q23

How often has each of the following happened to you during the past three months?

(PLEASE TICK ONE BOX ON EACH LINE)

Q23b

It has been difficult for me to fulfil my family responsibilities because of the amount of time I spent on my job.

0 Doesn't apply, no job

1 Several times a week

2 Several times a month

3 Once or twice

4 Never

8 Don't know, BG: can't choose, KR: don't know, refused

9 No answer

Missing information may actually be important for your analysis because it can signal an underlying problem with a specific question. Imagine if most or all of your participants skipped a question about a sensitive issue. Rather than being a nuisance, this would reveal significant information and it is therefore crucial to notate the missing information. One possibility is to leave the variable blank. Another is to put a specific number that is not included in the standardized variable code, such as 9999 or another eye-catching notation of your choice.

To summarize, when performing data entry, you must code all the variables, provide a unique identification number for each participant, create a detailed codebook, and decide on how to identify missing answers in the dataset. An important admonition is to be sure to save your data repeatedly as you work. Since data entry requires time and attention to details, a researcher needs to ensure that the work is repeatedly saved and backed up on other devices. Consider saving your work on various computers or a separate drive and emailing the file regularly to yourself or a colleague. Additionally, having a good work ethic and organization skills are crucial to this part of the research process.

ORGANIZING AND INPUTTING VARIABLES

As previously mentioned, each row represents one case or participant, and every column represents one variable in the same category for every single case or participant. We can look at a column and know that it refers to the same characteristic for each participant in the dataset. For example, the variable of the number of children participants have may be visualized as a long column of 0s, 1s, 2s, 3s, 4s, and 5s, where 0 represents no children; 1 represents one child; 2, 3, and 4 represent respectively two, three, and four children; and 5 represents five or more children. The same logic is applied to all the other variables.

However, each variable needs to be identified because it can be easily confused with the variable next to it. Every column calls for a descriptive name, just as each participant has an ID number. There are best practices when naming the variables in your dataset. Typically, the names are letters from the alphabet; special symbols or characters should be avoided. The data that you are entering in one software program may be later converted to a program that doesn't recognize the same symbols. The name of a variable is a single word, but may represent a combination of words, such as "raceethnic" for race and ethnicity, "socioecon" for socioeconomic status, or "healthpercep" for health perceptions. Numbers may be part of the variable name as well, but, ideally, these will be incorporated into a single word, such as "employm1" for employment 1 (i.e., primary work position), "diabetes2child" for children with Type 2 diabetes, or similar combinations. The length of the variable should be short, with a maximum length of 30 to 40 characters.

Variable Descriptions

If you are entering your data directly into SPSS or SAS, there will be boxes designated for a short description of your variables. This provides the information immediately within the dataset, without having to transfer information between files, but if you are using Excel, you may want to consider using an additional Word file to log these descriptive records. The description should include information on how the variable was measured, a list of the response choices, and other significant notes. For example, a researcher measuring acculturation of immigrants in the United States would create a set of questions to measure a person's degree of acculturation, where each question is translated into a variable at the time of data entry. In a brief description about what the variable is and exactly what it measures, the researcher may also note that it is part of a set of questions designed to measure acculturation.

Variable Types

After providing variable descriptions, we will look at variables from the perspective of how they have been measured. We will divide them into (1) nominal variables, (2) ordinal variables, and (3) numerical variables.

To recall from earlier in this chapter, one simple way to distinguish among these types of variables is to look at the answer choices. If the answers have no numerical value and no relationship to one another, the variable is nominal. If the responses do not have a numeric value, but represent ordered sequences of the responses, the variable is ordinal. Finally, when answers have mathematical meaning, they are called numerical variables. The numbers are numerically representational to the extent that we can calculate a meaningful mathematical average, subtract, and add, depending on the situation.

Notations for Responses

Before the work of data entry begins, the researcher needs to identify the notation for each answer choice. SPSS, SAS, Stata, R, M-plus, and other statistical software packages have a place to store information about the answer choices. If you are entering data into an Excel spreadsheet, it is a good idea to store your notations separately. The information reminds us at any given moment what the numbers in a variable column stand for. When we have many variables in front of us, we may forget what each represents—such as whether the number 5 in the "marstatus" variable means widowed, divorced, separated, single, married, or cohabiting.

Scales

We use psychometric scales as a universally accepted measure that ensures accuracy and avoids subjectivity. For example, instead of asking a participant to state whether his or her income is substantial or not, we provide income ranges to select. This avoids any judgments regarding level of income. Many researchers are passionate about finding the right instrument that will accurately measure their variables. This often relates to difficult concepts, such as the strength of the relationship between parent and children, acculturation level of the first or second generation of immigrants, level of addiction from some substance, and other subjective definitions.

When researchers attempt to measure a topic of interest, they may have their own ideas regarding what questions will best capture their intended meaning. It is practical, however, to have some scientific agreement on the instruments that can measure widely used concepts, such as the ones mentioned previously. A good number of studies must be conducted in order to create sturdy psychometric scales that can later be used by other scientists. A solid reference on different types of scales and their classifications can be found in the *Handbook of Tests and Measurement in Education and the Social Sciences* by Lester, Inman, and Bishop (2014).

It often happens that a single dataset includes many different measures or scales around areas such as likelihood or agreement. These same scales may measure a variety of concepts, be used for different variables in the dataset, and be organized differently. We strive for uniformity in the scales, but mistakes can happen. For instance, do some questions start with "strongly agree" and end with "strongly disagree," where others are in the reverse order? It is important to peruse the entire dataset a final time to make sure the various scales are consistently constructed. All the scales should be consistent, even if the concepts are unrelated. If we practice consistency, it can be especially helpful for the later stages of analysis when we may seek to portray information visually, for example, in bar charts, line graphs, or other types of figures.

Researchers sometimes group "strongly agree" with "agree" into one code and "strongly disagree" and "disagree" in another. They have the freedom to do so if the scale has four levels, with no neutral answer in the middle. Depending on the circumstances surrounding the study, this may offer the researcher more conclusive information.

Summary

Entering data into a spreadsheet is the first stage of analyzing your raw information. A logical format for data entry can be very helpful later on. Most researchers use a spreadsheet to enter the data so the information is accessible at any time in a format that is easy to analyze. The process of data entry starts by giving each participant a unique identification number.

We can identify two basic types of variables and four different subtypes: (1) categorical variables and (2) numerical variables. Categorical variables are further divided into (1) nominal variables and (2) ordinal variables, whereas numerical variables are divided into (1) ratio and (2) interval.

Nominal variables include information expressed in names, also known as qualitative variables. These variables can identify information such as gender, religious affiliation, race and ethnicity, political affiliation, marital status, and many other characteristics. Ordinal variables are usually measured in relation to one another. Some examples of ordinal variables are psychometric scales like "strongly agree" to "strongly disagree" or "very likely" to "very unlikely" about a topic of interest.

Numerical variables, also known as quantitative variables, are measured in numbers, such as age, income, and cost of home. There are two types of such variables: (1) ratio and (2) interval. The main difference between the two is based on the notion of the true zero point. Ratio variables have a true zero point whereas interval variables do not. Understanding the types of variables we have in our dataset paves the way for our analysis.

Once all the entries are identified, the coding begins. Coding refers to our ability to translate raw information into a form that is simple and useful for analysis. The specific way we code participant responses is saved in a separate file that we call a codebook. It compiles all of the information on the codes we created that accurately represents participant responses. When we code, we make sure to identify missing information and answers like "don't know" or "other."

The information is entered as a variable, and every variable needs an identifying name for its category. Horizontally, we look at each participant in our dataset, where one row shows one case or the information from one participant. Vertically, or in a column, we can see all responses for each variable. Each column represents one variable of information on something specific like race and/or ethnicity.

When we enter data in a file, we must include descriptive information about each variable. Sometimes information on what questions were asked or how the variable was coded can be lost, so it is crucial to keep all the coding information in separate and secure files. When we enter variables in a dataset, we recognize that they have different qualities, which also needs to be included in the descriptive notation. Since different variables are used for different types of analyses, the information on type of variables used is crucial.

Key Terms

Categorical variable: also known as a qualitative variable; a variable with answers that can be organized around various, often limited, categories.

Codebook: a document that lists the codes in a study and how they correlate with participant responses.

Interval variable: a variable that cannot express a ratio between the numbers because of the lack of a true zero.

Numerical variable: also known as a quantitative variable; a variable that is expressed with numerical values in the original data, such as age, income, or height.

Ordinal variable: responses that do not have a numerical value, but instead represent an ordered sequence of responses.

Ratio variable: a variable that can express a ratio between numbers because it has a true zero.

Taking a Step Further

1. Please identify the types of measurements (nominal, ordinal, ratio, or interval) in the examples provided below:
   a. Gender? (male/female)
   b. How often do you attend cultural festivals in your own community? (never/rarely/sometimes/often/always)
   c. Are you able to work in the United States? (yes/no)
   d. What is your height? (inches)
   e. Temperature over a year in a specific county? (F)
   f. How many hours do you watch television in a typical day? (number of hours)
   e. How much do you agree with the following statement: "Diversity is a benefit to our society's future"? (strongly agree/agree/neutral/disagree/strongly disagree)
   f. In what state were you born? (name of state)
2. Why is a codebook necessary when entering data in a spreadsheet?
3. What is the difference between qualitative and quantitative variables?
4. How can we handle missing data when entering the information in a spreadsheet?
5. How do we distinguish between ratio and interval variables?
6. Why do we include a description of our variables when entering data?

ANALYZING QUANTITATIVE DATA

CHAPTER OUTLINE

WHAT WILL YOU LEARN TO DO?

1. Summarize why statistics are used in research methods
2. Conduct univariate analysis
3. Describe the various ways to graphically represent data
4. Explain the measures of central tendency
5. Conduct bivariate analysis

ANALYZING QUANTITATIVE DATA

The topic of quantitative data analysis covers a great deal of ground. This topic continues to grow as researchers discover new ways of examining data; the struggle to better understand large amounts of data and information may never cease. This chapter focuses on some of the terminology and basic concepts that are used to analyze quantitative data.

Scientists use a branch of mathematics called **statistics** to analyze large amounts of data. Statistics is a broad discipline in its own right, and experience demonstrates that people with no mathematical background at all can not only participate in statistical work, but contribute brilliantly to the field. Statistics uses mathematical logic to interpret phenomena and draw conclusions from accumulated data. It establishes relationships between and among variables and can even attempt to predict future behaviors or relationships. Predicting future behaviors and relationships has become a new exciting playground for data scientists because we have combined the three disciplines of mathematics, statistics, and computer programming to advance our understanding of data and information. This chapter provides an overview of univariate and bivariate relationships between variables.

In analyzing quantitative data, we need to always be attentive to the presentation of the analysis. A picture is often worth a thousand words, so tables, charts, and figures are key in the process.

UNIVARIATE ANALYSIS

Univariate analysis is the analysis of only one variable. It is often descriptive in its nature because of the simplicity of focusing on a single variable, but it can also be inferential. **Descriptive analysis** allows researchers to sketch the details of variables and gain familiarity with the sample. In practical terms, it gives us an idea of the descriptive characteristics of variables. We become familiar with the minimum value, maximum value, range, quartiles, mean, mode, median (these definitions are explained later in this chapter), outliers, number of missing answers, and develop an understanding of the variable's distribution.

Descriptive statistics describe the first-time impressions of a variable. For instance, when we see someone or something for the very first time, we focus on superficial details. We immediately notice a person's hair length, hair color, weight, height, eye color, style of clothing, and other descriptive characteristics. Similarly, descriptive statistics offers this kind of first-time familiarity with the variables in our data.

Inferential analysis or **inferential statistics** provides a deeper understanding of variables because it draws conclusions about the population from the sample at hand. Inferential statistics is the best tool researchers have to make approximate generalizations about the entire population. This type of analysis goes beyond the description of the data available and attempts to draw relationships or associations between two or more variables. Once a relationship is detected between variables, inferential statistics attempts to predict the possibility of applying the same relationship to the entire population of the study. Inferential statistics can also be used to investigate one variable or to conduct *univariate analysis*. Inferential statistics operate by trying to calculate the **random error** that occurs during the data collection and analysis. Random error is associated with the type of common mistakes or errors that happen randomly when data are collected or entered in a spreadsheet or during the analysis. In Chapter 6, we refer to this random error as **sampling error** because it occurs during the process of data collection. In other words, sampling error includes the mistakes that happen randomly during the collection of data by either small mistakes in how the data are reported when collected, mistakes that can happen when the data are entered in the spreadsheet, or even small errors when the data are transformed or recoded during the analysis. It is important to note here that inferential statistics only investigates the sampling error that can happen randomly. It is unable to identify a sampling error that is repeated constantly because of some biases or other problems in the study's design. For example, let us assume that we have a variable that measures the types of mediums favored by painters, where 1 = acrylic, 2 = oil, 3 = chalk, 4 = acrylic pastels, 5 = oil pastels, 6 = pencils, and 7 = watercolors. Once the variable is entered in the spreadsheet, we may decide to recode the variable in smaller groups, so we simply want to divide participants favoring acrylic versus oil colors. If we happen to make a mistake of recoding chalk as oil or watercolors as oil, inferential statistics procedures would not be able to detect such a mistake. In this case, we are having a serious construct error of coding chalk and watercolors as oil. This is not random error, and it is very unlikely for inferential statistics to identify it.

Univariate analysis is the first step researchers take to familiarize themselves with the population sample. Before we can move forward with defining relationships among variables or predicting behaviors, we need to study the data we have collected. Univariate analysis helps to organize variables and gives us a clearer picture of our sample population. It is in conducting this analysis that we may better understand our population's basic demographics, including race, gender, age, nativity, religious affiliation, education, and other characteristics we measured in the data collection stage. If we conduct descriptive univariate analysis, we get a sense of these characteristics in our sample. Further, if we conduct inferential univariate analysis, we can see how our sample compares to the entire population on these same basic characteristics. There are a number of approaches available to the researcher conducting univariate analysis, depending on the needs of the study, such as frequency distributions and measures of central tendency.

Frequency Distributions

Constructing a **frequency distribution** with data is a simple procedure that can illustrate key characteristics of participants in the sample. The first indicator is what we call absolute frequency, which reflects the number of times an event or instance has been repeated. For example, an absolute frequency of 34 in the category of high school graduates (Table 10.1) indicates that 34 participants in our sample have graduated from high school with no further educational degrees.

The absolute frequency may not tell us much about population as a whole, but it helps to paint a more accurate picture if we know, for example, that the 34 participants with a high school degree make up the largest segment of the population. In fact, if we divide 34 by 71, the total size of the population, we get approximately 0.48. By multiplying 0.48 and 100, we determine that the percentage of our sample that completed high school is approximately 48%. Assigning percentages to each category adds another layer of valuable information about the data. In the example above, we can confidently report that almost half of the population has at least a high school degree and this is a more accurate portrayal of the information compared to the raw numbers. Percentages are capable of zooming out

TABLE 10.1 ■ Absolute Frequency Table in SPSS

Level of Education	Absolute Frequency
Elementary Education	10
High School Degree	34
Some College	5
College Graduate	15
Graduate Degree	7
Total	71

TABLE 10.2 ■ Frequency Table, Including Percentage and Cumulative Percent in SPSS

Level of Education	Absolute Frequency	Percentage	Cumulative Percent
Elementary Education	10	14%	1.00
High School Degree	34	48%	0.860
Some College	5	7%	0.380
College Graduate	15	21%	0.310
Graduate Degree	7	10%	0.100
Total	71	100%	1.00

from the raw numbers, by giving us a deeper layer of information because it compares the raw numbers to the entire sampling population. The third column in Table 10.2 shows these percentages.

Another descriptive measure of frequency is **cumulative percent,** which shows the collective percentage of categories together. This measurement collects aggregate numbers one by one and groups categories together to display the collective percentage of two or more categories together. For example, Table 10.2 shows that 0.380 of participants have at least some college education. We can also see that 0.860 of participants have at least a high school degree. We gain this information from the cumulative percent that groups together categories one by one.

Another valuable feature of cumulative percent is its ability to provide cumulative percentages when multiplied by 100. Therefore, we can say that 38% of our sample has at least some college or higher and 86% of participants graduated high school or higher. Cumulative percent is sensitive to the order of organization. In our graduation example, imagine if we organized the cumulative percent from the bottom up instead of the top down. Our findings would be meaningless because we would say that 0.90 or 90% of the sample has a college degree or less. The cumulative percent needs to be logically organized to produce meaningful information.

In recent years, statisticians, mathematicians, and computer scientists have begun to collaborate in a new field called **data science**. Data science is the ability of a computer or a machine to think and make decisions based on previous data or information. By using predictions, statistical skills, and programming knowledge, we are able to analyze a large variety of data and use the results to improve our lives. If you are curious to learn more about data science, look at free introductory courses like the following course hosted by Udacity: https://www.udacity.com/course/intro-to-data-science--ud359.

GRAPHICAL REPRESENTATION OF DATA

Although tables like Tables 10.1 and 10.2 are meaningful for the researcher, a graphic representation of the information can have a powerful impact on a broader audience. A graph can visually depict relationships and patterns among variables. Some common graphs used in descriptive statistics are bar graphs, histograms, line graphs, and pie charts. Let us walk through the characteristics and uses of each kind.

Bar Graphs

Bar graphs use vertical bars that represent each variable in a study. A longer bar represents a higher frequency for a variable or category, whereas a shorter bar implies less frequency. The viewer can easily distinguish not only which category occurs most frequently, but how it compares to the other categories. For example, the bar graph in Figure 10.1 details a race and ethnicity variable. There are three categories within this single variable: (1) Caucasians, (2) African Americans, and (3) Hispanics. Bar graphs are often used for qualitative variables like these in which the bars display the frequency of separate and distinct categories. There is no relationship between the variables in this category; however, when portrayed in a bar graph, we can see that Caucasians have the highest frequency, followed by African Americans. We can also see that there are very few participants from the Hispanic population.

Histograms

Closely related to a bar graph, **histograms** are often used for quantitative variables where the categories are related to one other. A histogram presents the frequency of a variable in the form of distinct but connected bars. Since histograms are used for quantitative variables, the distribution takes on a meaningful profile because it compares each category with the rest of the data. It is like you are looking at the information from above and can see how each category relates (it is bigger or smaller) to all the other available categories. Look at the histogram in Figure 10.2, which shows the number of hours participants report they watch television in a regular workday. This variable was selected from the General Social Survey of 2013, which had 2,600 participants. The distribution depicted in Figure 10.2 shows that 1,669 participants responded to this question. Most participants reported watching 4 hours or less of television per day. We can also see that around 120 participants do not watch television at all, and approximately 310 only watch 1 hour per day. We can also note that less than 100 participants watch 8 hours or more of television per day.

FIGURE 10.1 ■ Race and Ethnicity Bar Graph

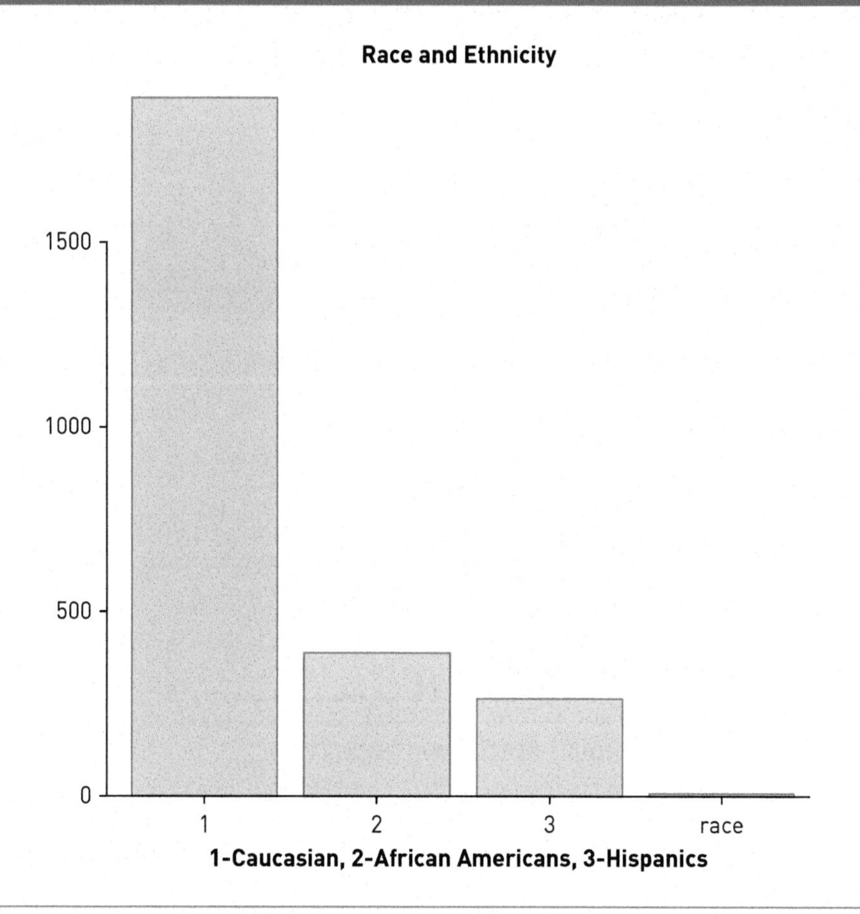

Source: From the General Social Survey 2013 dataset.

As you can see from Figure 10.2, there is no space between the histogram bars unless a category is not represented, such as participants who are not watching television at all. Each bar in the histogram shows the frequency of participants who reported watching the same number of hours of television per day.

Line Graphs

A line connecting points in a distribution is known as a **line graph**. Line graphs are useful in depicting two or more categories simultaneously. If there are multiple distributions

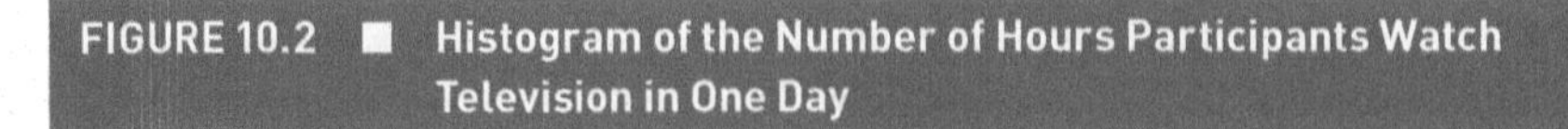

FIGURE 10.2 ■ Histogram of the Number of Hours Participants Watch Television in One Day

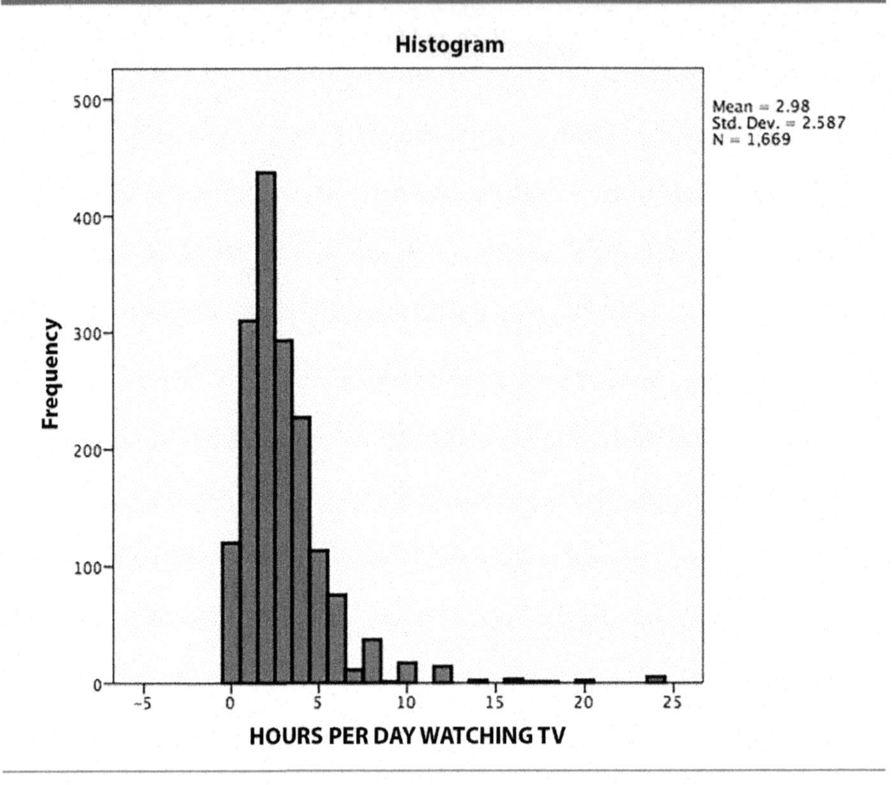

Source: From the General Social Survey 2013 in SPSS.

in the same category, showing them this way can be an effective choice. For example, we may want to measure the effectiveness of advertising on television (TV) and the sales of a particular product. The line graph in Figure 10.3 depicts the relationship between the amount of advertising on TV and the actual sales of the products being advertised (James, Witten, Hastie, & Tibshirani, 2009). Just by looking at the line graph, we can clearly see the effectiveness of advertising because the more one product is advertised, the more it sells (the lines are progressively reaching the very right end of the coordinate system). In other words, a line graph is an effective way to demonstrate relationships between variables.

If you remember from your geometry courses, the horizontal axis is called abscissa, and the vertical axis is called ordinate. They are also called the x-axis and y-axis, respectively,

FIGURE 10.3 ■ The Relationship Between Advertising on TV and Sales Expressed in a Line Graph

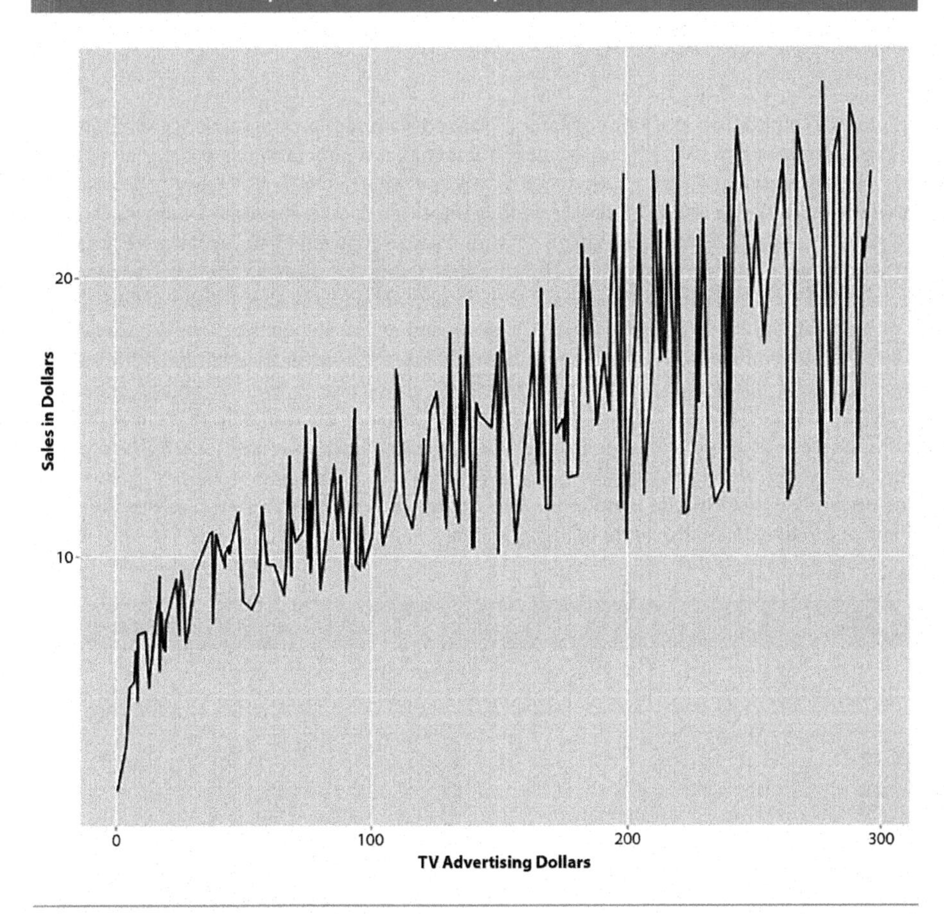

and are parts of the coordinate system. We use this coordinate system to determine the position of a point. The coordinate system has been widely used in other fields, including statistics. In Chapter 9, we discussed how to code answers from participants by assigning numbers to them. Once the answers are coded in numbers, we can easily transfer them in a coordinate system and look for relationships between variables. It is exciting to see the ability of scientists to use interdisciplinary skills and knowledge to determine relationships that are important in our daily lives.

ETHICAL CONSIDERATION 10.1

REPRESENTATION OF DATA

It may be tempting to manipulate the scales of graphs to make an impact on the audience when we present findings. The same graph in two different scales illustrates the information we have found in the data accurately so we are not bending any rules technically speaking. However, a larger coordinate system (like one that starts at 0 and ends at 1,000) can reduce the differences between the bars in a histogram or bar plot, at least visually. Simultaneously, a smaller coordinate system can emphasize these differences. Consider the two graphs shown below. They are both using the same variable of gender of 133 participants where 116 are women and 17 are men. The first graph is based on a scale of 0 to 120 since the largest group (women) is not bigger than 120. The second graph shows the same frequency, but on a scale of 0 to 300. As you can tell, the first graph emphasizes the difference between the two groups whereas the second graph minimizes this difference. When we are shown the second graph, we may think that the difference between the two groups is not as large as we would think if shown the first graph. Sometimes we are tempted to impress audiences and claim a bigger or a smaller difference than the results imply. Although we are not breaking any laws of statistics for doing so, we are inadvertently manipulating our audience. Although tempting to show otherwise, as researchers, we are bound by rules of ethics and our goal is to analyze data to the best of our abilities.

FIGURES 10.4 ■ Two Graphs on the Same Variable, but Different Scales

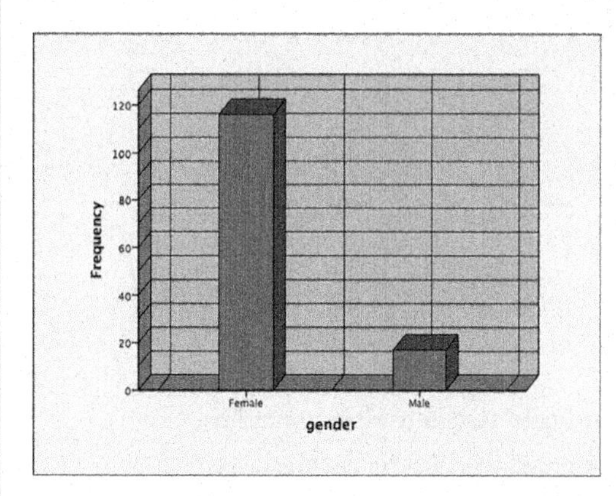

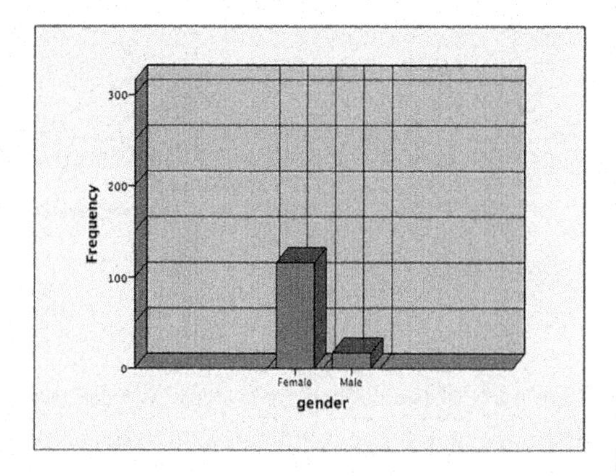

Pie Charts

Pie charts are also a visual representation of the distribution of one variable and work best with categorical variables that have a small number of categories. A researcher may have a categorical variable with many divisions that may not be well represented in a pie chart. For example, if we have a variable measuring religious affiliation where more than 10 different religious affiliations are categorized, our pie chart may not best serve our needs. It has the possibility of looking crowded and we may not make the point we want to make. In cases where the number of categories is relatively small and one category is much larger than the rest, a pie chart may truly help us present our data in the most effective way. The pie chart shown in Figure 10.5 depicts respondents' confidence in the education system. The data were taken from the General Social Survey in 2013 where 1,667 participants responded.

The main advantage of a pie chart is its visual impact on the audience. For example, this pie chart clearly shows that almost a quarter of the distribution has hardly any confidence in the education system, and only a quarter has a lot of confidence in the system. We can clearly see how the majority of people who participated (over 50%) are somewhere in the middle, having only some confidence in the system. This visual representation makes a major statement if we are researching the extent of peoples' confidence in the education system.

FIGURE 10.5 ■ Pie Chart Depicting Participants' Confidence in the Education System

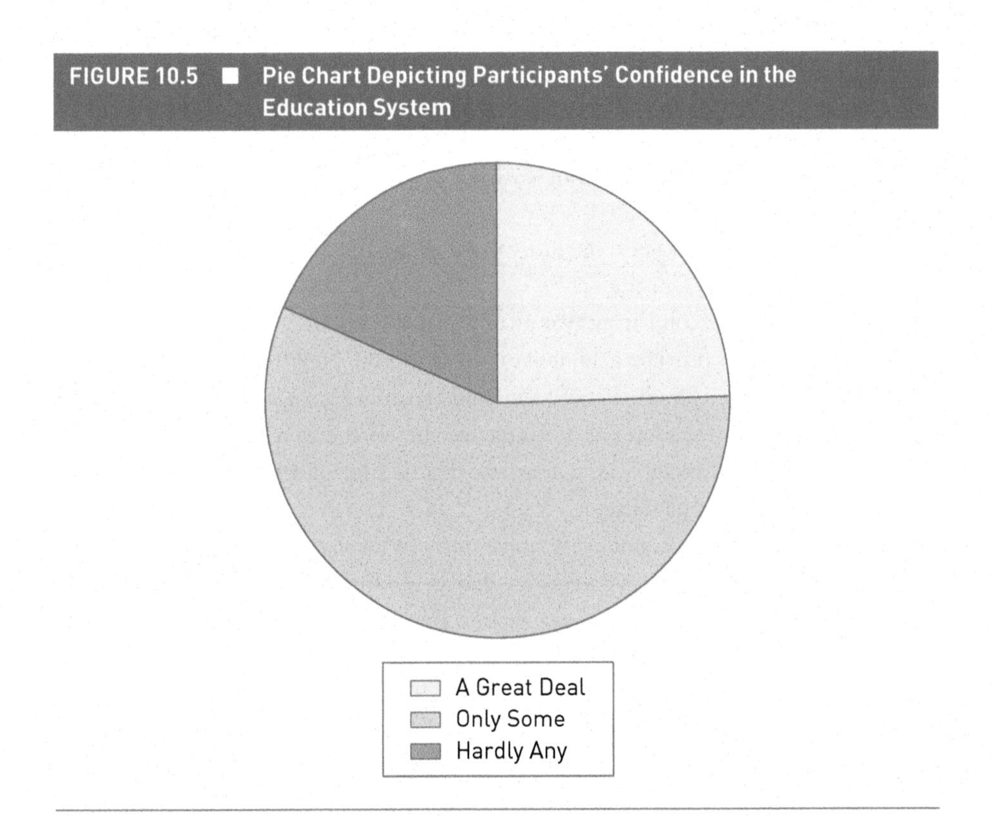

Source: From the General Social Survey 2013 in SPSS.

MEASURES OF CENTRAL TENDENCY

The **measures of central tendency** are widely used in statistics. These measures summarize an entire variable, or category, in a single value that is numerically representative of this variable. When we state that a nurse earns, on average, $69,000 a year, we are referring to the mathematical average of all nurse salaries, according to the U.S. Bureau of Labor Statistics in 2015. The **average** is a descriptive number that represents the variable of U.S. nurse salaries in a given year. The average is a widely used measure of central tendency called the **mean.** Other measures of central tendency include the **median,** or the middle point in a distribution; the **mode,** or the number that occurs with the highest frequency; the **range,** or the difference between the maximum and minimum values in a set of numbers; **quartiles,** or the division of a distribution in quarters; **standard deviation** and **variance** that measure how spread out a distribution is from the mean; and **outliers** or the information that is in the far end of a distribution. Let us take a brief look at each of these concepts.

Mean, Median, and Mode

The *mean* is the mathematical average of a set of numbers. All the numbers in a variable category are added together and then divided by *n,* or the total entries in that dataset. The widely used notation for the mean is x-bar.

x-bar = the sum of all numbers/the *n*

The mean is useful for comparing two or more groups, because the difference in their mathematical averages provides a point of easy comparison. Looking back at the example of nursing salaries, the U.S. Bureau of Labor and Statistics (2015) showed an average salary of $69,000. The same source also tells us that the average salary of a family doctor is $150,000. Given these averages, we can surmise that the average salary for a family doctor is more than twice that of a nurse.

The mean takes into account every single point in the variable. It is also very sensitive to outliers. An outlier is a data point that is too high or too low in comparison to the majority of other points. Going back to the salary example, let us assume that there is one nurse who makes $500,000 per year in our dataset. This nurse's salary can increase the mean of nurses' salaries and mislead us into thinking that the average salary of a family doctor and a nurse is similar when that is not true for the majority of cases.

The *median* represents the center point in a set of numbers that have been arranged in ascending or descending order. Let us take a look at a short list of data points:

2, 4, 6, 8, 20, 24, 30, 32, 34

In this set of numbers, the middle number, or median, is 20. If we were to add another number at the end of this set, it would result in an even number of points, as follows:

2, 4, 6, 8, 20, 24, 30, 32, 34, 40

The median is then calculated as the average of the two middle numbers. The median in this example would be 22 (because $20 + 24 = 44$; $44/2 = 22$). The median is very useful in statistics because it gives us an understanding about the distribution of a variable. Because the median is always positioned exactly in the middle of a distribution, we know that 50% of the distribution is below the median and 50% is above the median.

For example, if you are interested in buying a house and are looking at different neighborhoods in an area, the information about the median gives you an understanding of house prices in this neighborhood. If you know that the median price is $150,000, you understand that 50% of the houses in this neighborhood will cost less than $150,000 and 50% of the houses will cost more.

The *mode* is the most frequently used number in a dataset. It tells us which number is most commonly encountered or has the highest frequency in a specific set of numbers. Let us look at an example of a set:

2, 4, 6, 6, 6, 6, 6, 8, 10

In this example, the frequency that is most commonly encountered, or the mode, in this set is 6. If we were to add the number 40 in this same set or the number 100, the mode would still be 6. This means that the mode is somewhat resistant to additional numbers in the distribution (even outliers). This quality makes the mode a preferred source of information when the variable includes outliers. If the distribution includes outliers, the mean may not be representative of the distribution and even the median (although less affected) may be misleading. The mode, on the other hand, can provide additional information that can potentially clarify any misleading information from the first two measures.

Taking the house price example from before, if we know that the median is $150,000, the mean house price in the neighborhood is $200,000, and the mode is $100,000, we can clearly understand the presence of some outliers (i.e., houses that cost a lot in this same neighborhood that have artificially increased the mean). Now we know that the most frequent house buyer in this neighborhood only pays $100,000. Having all three measurements of central tendency can give us a clearer picture of the distribution of a single variable. Finally, it is important to mention that one variable can have more than one mode. Often a distribution is such that there is more than one number with high frequency. When a variable has only one mode, it is called *unimodal*, when a distribution has two modes is called *bimodal*, and when a distribution has three or more modes, it is called *multimodal*.

Variability

Variability expresses the degree to which data in a dataset spreads out, or deviates, from the center. At the center of a dataset are our measures of central tendency: the mean, mode, and median. Some datasets include points that are distant from the mean, while others reside close to the center. When the numbers greatly deviate from the center value, the dataset is said to have a lot of variability. Let us put this into context so the importance of variability in our analysis becomes clear.

The following two sets of numbers (variables) show the ages of participants in two different studies.

Variable A

20, 22, 24, 26, 28, 22, 24, 24, 26, 20, 20, 26, 26, 24, 28, 22, 28, 22

Variable B

15, 16, 0, 16, 18, 24, 30, 36, 35, 14, 37, 38, 14, 42, 12, 30, 40, 15

Eighteen participants have volunteered for each of these studies; the average age for each participant set is 24. This average can mislead us to think that the participants' ages are similar between the two variables if we do not consider the variability in each dataset. Despite the variables' equivalent means, these sets are divergent in nature. Variable A has very little variability compared to Variable B. Can you see these differences in variability by scanning these sets of numbers?

In the first set, the ages are closely clustered around the center, ranging from 20 to 28. In the second set, the ages are widely dispersed, from 0 to 42. When sets have little variability, we consider them to be *homogenous,* whereas a lot of variability is described as *heterogeneous.* While we love diversity and variety in our daily life, we are not so thrilled to encounter heterogeneous variables. For analytical purposes, heterogeneous variables pose a higher level of complexity in approaching our analysis. How can we measure the variability of complex sets composed of numerous participants when it is not possible to simply scan the numbers? This is where *range, interquartile range, outliers, variance,* and *standard deviation* can help.

Ranges

Range is the difference between the largest and the smallest numbers in the set. In the previous example, we looked at the age ranges of two sets of participants: one set of numbers ranged from 20 to 28 and the other one ranged from 12 to 42. In the first variable, the range is 8, whereas in the second the range is 30. A larger range reflects more variability in the dataset than a smaller one. Mathematically, we express range with the following formula:

$$\text{Range} = \text{Maximum value} - \text{Minimum value}$$

Interquartile range (IQR) is another measurement of the middle 50% of the scores in a variable. To find the IQR, we divide the variable into chunks of 25% each, where the first 25% is Q1, the second 25% that falls right in the midpoint or the median of the distribution is Q2, and the upper 25% is Q3. IQR is then calculated as the difference between Q3 and Q1, so IQR = Q3–Q1.

The interquartile range is used to define whether we have some outliers in our variable. Researchers have agreed (Jaccard & Becker, 2002) that any score that is higher than Q3 + 1.5*IQR or lower than Q1–1.5*IQR is considered an outlier. An explanation of outliers and their role in data analysis is provided at the end of this chapter. Let us look at a simple example to see the power of IQR in action. Consider the following data points that make up a variable (ordered from lowest to highest). This variable is illustrated in Figure 10.6 in a boxplot where the outliers are visible:

1, 4, 6, 7, 8, 8, 8, 8, 9, 9, 10, 11, 13, 13, 40

There are 15 scores in this set of numbers. The median will be the eighth number in the set, or 8, so Q2 = 8. We know that 50% (seven numbers) are below and 50% (seven numbers) are above this Q2. If we divide each part by half to find each quarter (25%), we can see that Q1 is the fourth number (7) and Q3 is the twelfth number (11). Now, following the formula for IQR: Q3–Q1 = IQR, 11–7 = 4. The IQR = 4.

Using calculations, we can determine if there are any outliers in this set of numbers. The lowest outlier formula is Q1 – 1.5*IQR, which means 7 – 1.5*4 = 7 – 6 = 1. Any number lower than 1 will be considered an outlier for this set of numbers. The set starts at 1 and

FIGURE 10.6 ■ Boxplot of IQR Illustrating the Outlier as Well as the Upper and Lower Limits

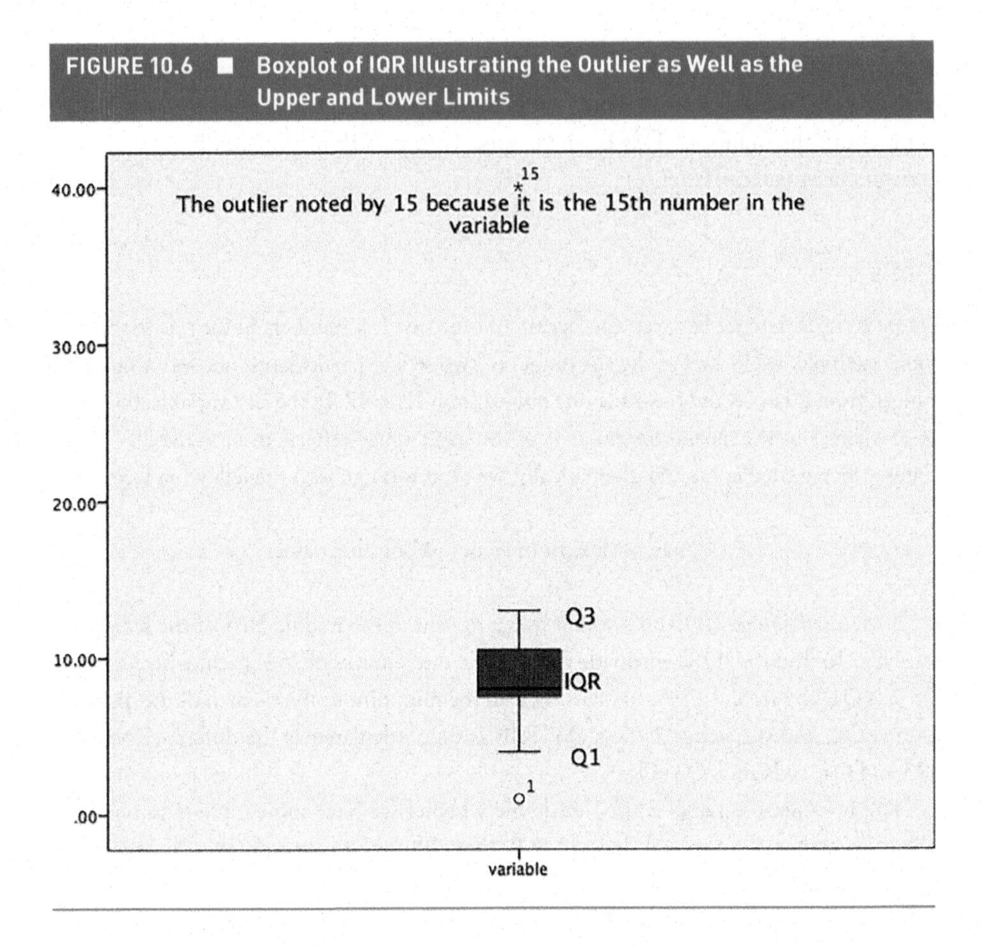

there are no numbers lower than that, so we can state that there are no lower outliers. For the upper outlier formula, Q3 + 1.5*IQR, we calculate 11 + 1.5*4 = 11 + 6 = 17. Therefore, any number higher than 17 in the set will be considered an outlier. Forty is clearly higher than 17, therefore it is an outlier for this set of numbers.

Measures of Variability

Variance and *standard deviation* are commonly used and closely related measures of variability. The standard deviation is the square root of the variance. The variance and the standard deviation represent each data point's average deviation from the mean. Variance is the average sum of squared distance of every single point in the dataset from the mean. The standard deviation is the square root of the variance, also representing the average distance of every point from the mean. The example below gives an understanding of how variance and standard deviation are calculated. Both standard deviation and variance are useful because they consider all the numbers, or values, in the set and provide an accurate information of how spread out are the data in the entire variable. To better understand each of these concepts, we need to examine how they are measured.

Table 10.3 reiterates the participant age values that we used in Variable A. First, we calculate the difference between every single value (number) in the set and its mean. These values are found in the second column. If we were to add these numbers together, the result will always be 0.

The mean of the set is the mathematical average of that set. Every other number in the set will be higher, lower, or exactly equal to the mathematical mean. This is a simple mathematical rule regarding the mean. Since we are subtracting the mean from each

TABLE 10.3 ■ Calculating Variance and Standard Deviation

Variable A	(x-bar)	(x-bar)^2
20	-4	16
22	-2	4
24	0	0
26	2	4
28	4	16
22	-2	4
24	0	0
24	0	0
26	2	4
20	-4	16
20	-4	16
26	2	4
26	2	4
24	0	0
28	4	16
22	-2	4
28	4	16
22	-2	4
Sums $\Sigma = 432$	$\Sigma = 0$	$\Sigma = 128$

number in the variable, some numbers will be negative, some numbers will be positive, and some numbers will be 0. The rule for the mean tells us that the sum of these differences will always equal 0 for any variable used. Therefore, we must square these differences from the mean (as shown in column 3) to remove the negatives and avoid a 0 result. The sum of the squared differences of each value from the mean gives us the number 128. If we then divide this sum by the number of participants in the study (or by the number of values we have in the set), we get an average distance from the mean of all the scores in the variable. We call this number a variance, and in this case, it equals around 7.11.

The formula for the variance is

$$\text{Variance} = \Sigma(\text{x-bar})^2/N$$

Recall that we squared all the differences of each number from the mean. Therefore, the square root of the variance is what we call the *standard deviation*. The formula for the standard deviation is

Standard deviation = square root of variance [expressed in the formula here]

In this example, it equals approximately 2.66. The standard deviation is widely used in research to tell us the average distance of the scores from the mean. When we write that the average income of a certain neighborhood is $45,000 per year with a standard deviation of $2,000, we are implying that the average amount of every family's income from the mean income is $2,000. In practice, we understand that the family income of the majority of participants in the dataset varies from $43,000 to $47,000 per year.

Normal Distribution

Normal distribution and *z-score* are two other important concepts in statistics that relate to the spread of a dataset, or its variability. The normal distribution is a perfect theoretical distribution where the mean, the mode, and the median are equal to one other. It is symmetrical as the mean divides the distribution into equal halves. Therefore, 50% of the values in the distribution are less than the mean, and 50% of the values in the distribution are greater than the mean. If we were to visualize the normal distribution (see Figure 10.7), it would look like a symmetrical bell curve where the mean, median, and mode all align at the same point in the center.

You are probably wondering why statisticians spend time thinking about a *theoretical* distribution like the normal distribution rather than focusing on real distributions. However, you may be surprised to learn how useful the normal distribution can be in analyzing data. One widely used property of the normal distribution is that it is divided proportionally between *standard scores,* also called *z scores.* Standard scores indicate the

FIGURE 10.7 ■ Normal Distribution in Python 2.7

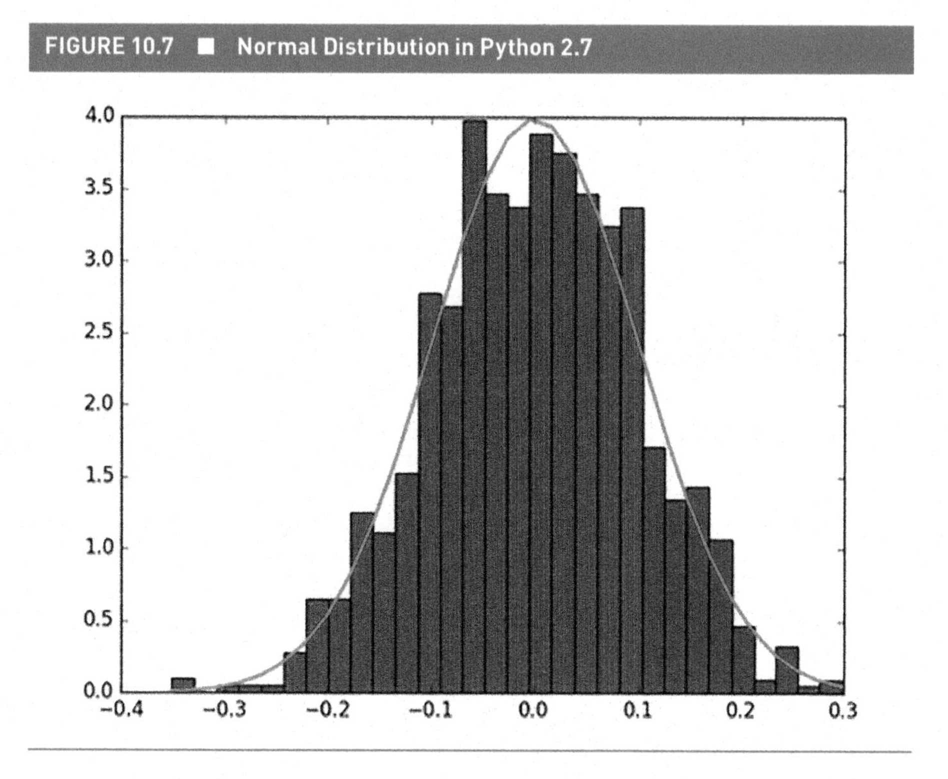

number of standard deviations in a distribution. A standard deviation may have a value of 10, for example, but a standard score of 2 indicates that there are 2 standard deviations in this distribution (0.2 and −0.2 in Figure 10.7).

We need standard scores to tell us how many standard deviations are found in the distribution and what proportion of the scores fall within this range. When considering the normal distribution in Figure 10.7, the mean, the median, and the mode all fall at 0 in the middle of the distribution. The right side is divided by three scores (0.1, 0.2, and 0.3), and the left side is also divided by negative scores (−0.1, −0.2, and −0.3). They correspond to 1, 2, and 3 and −1, −2, and −3 standard scores, respectively. The proportion of scores that fall between the standard scores in any normal distribution, regardless of its means or standard deviations, are defined by the following rules:

1. The area between 0 and −1 and the area between 0 and 1 always equals .3413.
2. The area between 1 and 2 and the area between −1 and −2 always equals .136.
3. The area between 2 and 3 and the area between −2 and −3 always equals .0215.
4. The proportion of scores greater than 3 or smaller than −3 is .0013.

FIGURE 10.8 ■ Normal Distribution

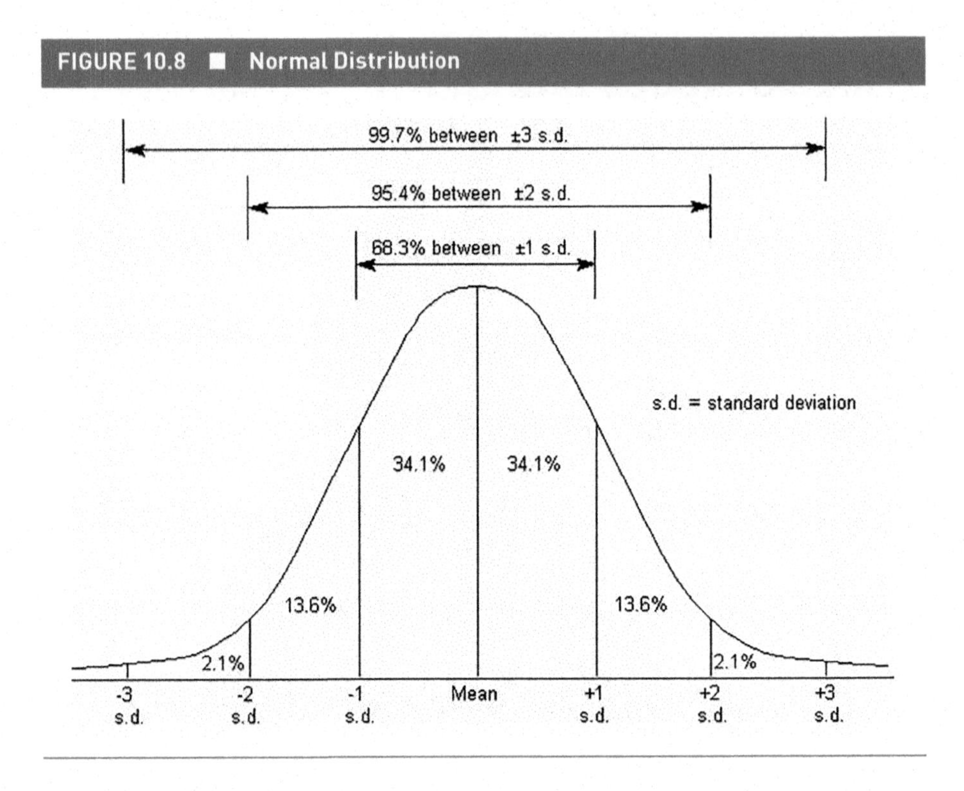

This information is important because it means that the area between −1 and 1 equals .3413 + .3413 = .6826. Therefore, we can conclude that approximately 68% of the distribution falls between a standard score of −1 and 1. In other words, 68% falls between one standard deviation below and above the mean. Furthermore, the distribution between −2 to 2 z scores or 2 standard deviations below and above the mean is .9974 or almost 99% of the data values. It is desirable to have our dataset approximate a normal distribution because we are then able to draw conclusions about variability, such as that most of the values, or about 68%, will be between ± 1 z score, and so on.

Sometimes, we may need to find the standard scores for a specific distribution. To do so, we follow the formula for standard score based on the raw scores. This formula is

$$(x - mu)/sd$$

x is a point in the distribution,

mu is the average

sd is the standard deviation.

Skewed Distributions

Many distributions do not behave like normal distributions. They lack the perfect symmetry of having the mean, median, and mode in the center and are asymmetric in their shape when viewed in a histogram. We call this **skewness**. Skewness in a distribution can be on the right or the left side depending on the quality of the data. A left skewed distribution has a long tail on the left side and the majority of the data are on the right. This is called a left-skew, left-tailed, or negatively skewed distribution. A distribution with a long tail on the right but with the majority of points on the left is called a right-skew, right-tailed, or positively skewed distribution.

For example, an open source community where you can learn programming and coding called Free Code Camp has made a large amount of data available about the people who are learning to code. They have surveyed 15,000 people who are actively learning coding and the data depicted in the histogram in Figure 10.9 are taken from their raw file of information that they have made available to the public. This histogram shows that the majority of people who are learning to code are younger than 40 years old. The distribution is right skewed or positively skewed. If the majority of participants would have been older than 40, we would have a negatively skewed or left-skewed distribution.

FIGURE 10.9 ■ 2016 Free Code Camp Survey Database: Ages of 15,000 Participants

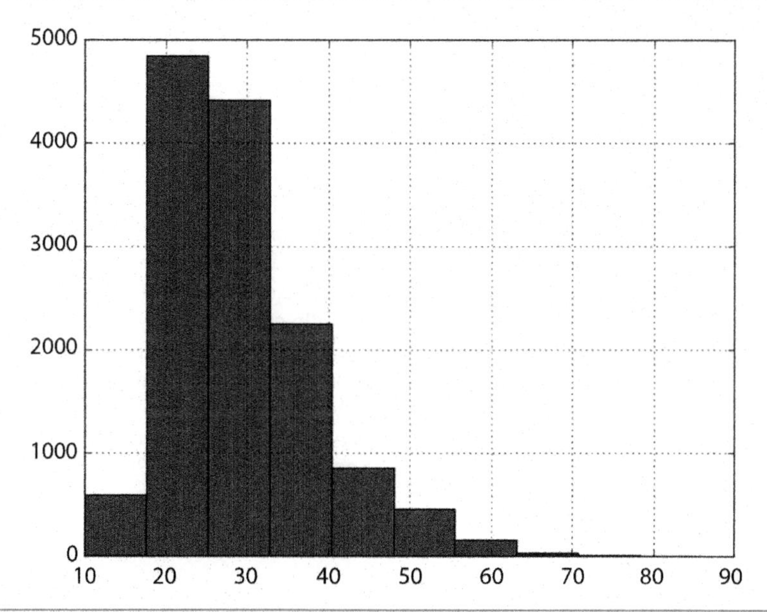

RESEARCH IN ACTION 10.1

SAMPLE SIZE AND SKEWNESS

Source: Piovesana, A & Senior, G. (2016) How Small Is Big: Sample Size and Skewness, *Assessment*, DOI: 10.1177/1073191116669784. http://journals.sagepub.com/doi/full/10.1177/1073191116669784

Sample sizes of 50 have been cited as sufficient to obtain stable means and standard deviations in normative test data.

The very first sentence tells us that a sample size of 50 is generally considered sufficient to get a normal distribution. Note how the authors are talking about "normative test data," not just any data. The term "normative test data" refers to the data that are used to portray the usual or the defining characteristics in a population (O'Connor, 1990). These are data used to describe rather than explain relationships of variables and are extremely important to help researchers and clinicians to understand what is "common" and "usual" in a specific population.

The influence of skewness on this minimum number, however, has not been evaluated. Normative test data with varying levels of skewness were compiled for 12 measures from 7 tests collected as part of ongoing normative studies in Brisbane, Australia. Means and standard deviations were computed from sample sizes of 10 to 100 drawn with replacement from larger samples of 272 to 973 cases. The minimum sample size was determined by the number at which both mean and standard deviation estimates remained within the 90% confidence intervals surrounding the population estimates.

The authors of this article are examining the effect of skewness on small normative samples. They have compiled 12 samples measured from seven different tests of skewed data to see the effect of skewness in small sample sizes. They have calculated standard deviations and means from different sample sizes to compare their findings from 10, the smallest, and 973, the largest. They are also reporting using a confidence interval of 90%.

Sample sizes of greater than 85 were found to generate stable means and standard deviations regardless of the level of skewness, with smaller samples required in skewed distributions. A formula was derived to compute recommended sample size at differing levels of skewness.

Their findings indicate that a sample size of 85 or larger is sufficient to generate stable means and standard deviations regardless of the amount of skewness. They also report the formula in the article. We can conclude from the abstract that a sample size of 50 (when skewness is present) may not be sufficient for reliable means and standard deviations. The authors are indicating that a sample size of at least 85 participants would be a much more reliable size in normative studies.

What does it mean for the central tendency if a distribution is skewed on the right or the left? Remember that in a normal distribution, the mean, median, and mode are all on the same point. If the distribution is skewed to the right (positively skewed), the mode will be on the far left of the distribution where the highest peak is and the median will be farther right from the mode (so the median will have a greater value than the mode). Finally, the mean will be the farthest on the right of the distribution with an even higher value than the median and mode. This is an important characteristic for any distribution that is being examined.

The opposite occurs in a negatively skewed (left-skew) distribution. In this case, the mean will be on the far left and have the smallest value, followed by the median, which will have a greater value. Finally, the mode will be the farthest on the right, where the highest peak of the distribution is located.

Continuous and Discrete Variables

As you may recall from Chapter 9, a categorical variable does not take all possible units and has distinct numbers that refer to specific categories (such as 1, 2, 3, and so on). For example, if we were measuring how many children respondents have, their answers would be whole numbers because it is not possible to have 1.5 children. These categorical variables that can only take whole numbers and do not have any continuity in terms of measurements are called **discrete variables**. A **continuous number**, on the other hand, is a type of numerical variable that can take all possible units, such as a measure of income, height, and weight. We can make $51,250.24 per year or be 6.2 feet tall. It is important to put that knowledge of the types of variables in context of the available statistical analyses.

Some of the measures described up to this point do not apply to discrete variables, and others may not provide useful information about the data. For example, if we were to measure religious affiliation with responses, such as 1 for Catholic, 2 for Protestant, 3 for Muslim, and 4 for Jewish, it would not be useful to measure the mean, median, or standard deviation using these responses. An average of 2.5 would not make any logical sense. However, we can use the mode, for instance, to establish that most responses are 4—the majority of the respondents are Jewish.

The mean, mode, median, and standard deviation are useful in the case of continuous variables, or variables with values that increase by specific units, such as height, weight, age, income, temperature, blood pressure, and daily caloric intake. It would make sense, for instance, to state that the average age of the sample is 24 years old, with a standard deviation of 3. Following the logic regarding normal distributions, this would mean that 68% of the people in the sample are between the ages of 21 and 27.

BIVARIATE ANALYSIS

Bivariate analysis is used to investigate the relationship between two variables. In particular, we are interested in understanding whether the changes in one variable are reflected by changes to another variable. Although there are many different ways that two variables may relate to each other, social statistics offers insights into understanding the relationship between two quantitative variables via **correlation** and **regression**. Correlation can be used to determine the extent to which two quantitative variables approximate a linear relationship. Regression can be used to identify the line that, though imperfect, best describes this relationship (Jaccard & Becker, 2002, p. 136). Therefore, if we were to enter two quantitative variables in a scatterplot and look at them, we should be able to identify whether the connecting dots are scattered or whether they approximate a line. If these dots seem to be drawing a line, we are looking at a linear relationship between variables that are correlated to each other.

The Linear Model

A linear relationship between variables is a perfect relationship between two variables that, when graphed, resembles a straight line. In a linear relationship, the changes in units of one variable are reflected in specific interval changes in the other variable. Linear relationships exist mathematically, but they are very rare in research findings because most of the variables we collect from participants will have some variability in the answers. This is mostly true for social and behavioral sciences because our participants are often humans. As you already know, humans are often unique in their life experiences and preferences, so it is difficult to have their answers appear as perfect linear relationships. If we were to conduct a study about a specific variety of flowers, for example, and look at the relationship between the petal lengths and petal widths—as is the case of the widely used free dataset of Iris (you can find the entire Iris dataset at the UCI dataset repository: https://archive.ics.uci.edu/ml/datasets.html), we will get a nearly perfect linear relationship. Here is another such perfect linear relationship. Table 10.4 shows the number of tomato plants purchased by 20 individuals and the amount they paid for the plants. Each tomato plant costs $10, so the more plants an individual purchased, the greater the amount he or she paid.

TABLE 10.4 ■ Tomato Plants and Amount Paid

ID Buyers	Tomato Plants	Amount Paid ($)
1	1	10
2	2	20
3	3	30
4	4	40
5	5	50
6	6	60
7	7	70
8	8	80
9	9	90
10	10	100
11	11	110
12	12	120
13	13	130
14	14	140

(Continued)

TABLE 10.4 ■ (Continued)

ID Buyers	Tomato Plants	Amount Paid ($)
15	15	150
16	16	160
17	17	170
18	18	180
19	19	190
20	20	200

A scatterplot with this information will resemble 20 connected points that are part of the same straight line (see Figure 10.10). If we connect the points, we can see the linearity of the relationship between these two variables (i.e., the number of tomato plants purchased and the amount of money paid). The points in the first figure represent the relationship between these two variables rather than the line. The line is imaginary, because not every single point in the line reflects a meaningful relationship. No one can buy 1.5 tomato plants and pay $15. If we were to express this relationship mathematically, we would say that $y = 10x$, where y is the amount of money paid and x is the number of plants purchased. The formula $y = 10x$ expresses the relationship between the number of plants purchased and the cost of the plants. The number 10 tells us the units by which the y changes every time x changes. In statistics, this number is called the *slope.* The formula expressing a linear relationship is $y = bx$, where b is the slope.

When the slope is a positive number, the linear relationship between variables is also a *positive relationship,* and looks like the example illustrated. When the slope is a negative number, the linear relationship is a *negative relationship* or inverse relationship.

Correlation

When all the data are cleaned and entered in a software program, scientists will typically run an analysis using a *Pearson correlation coefficient,* called r—an index that determines the level of correlation between two variables. The mathematics of correlation are beyond the scope of this textbook, but the means of interpretation is important to understand the analysis.

Statistical programs will often give us the correlation coefficient and by examining it, we can have insight into the relationship between two variables. Pearson correlation

FIGURE 10.10 ■ The Relationship Between Tomato Plants Purchased and Cost

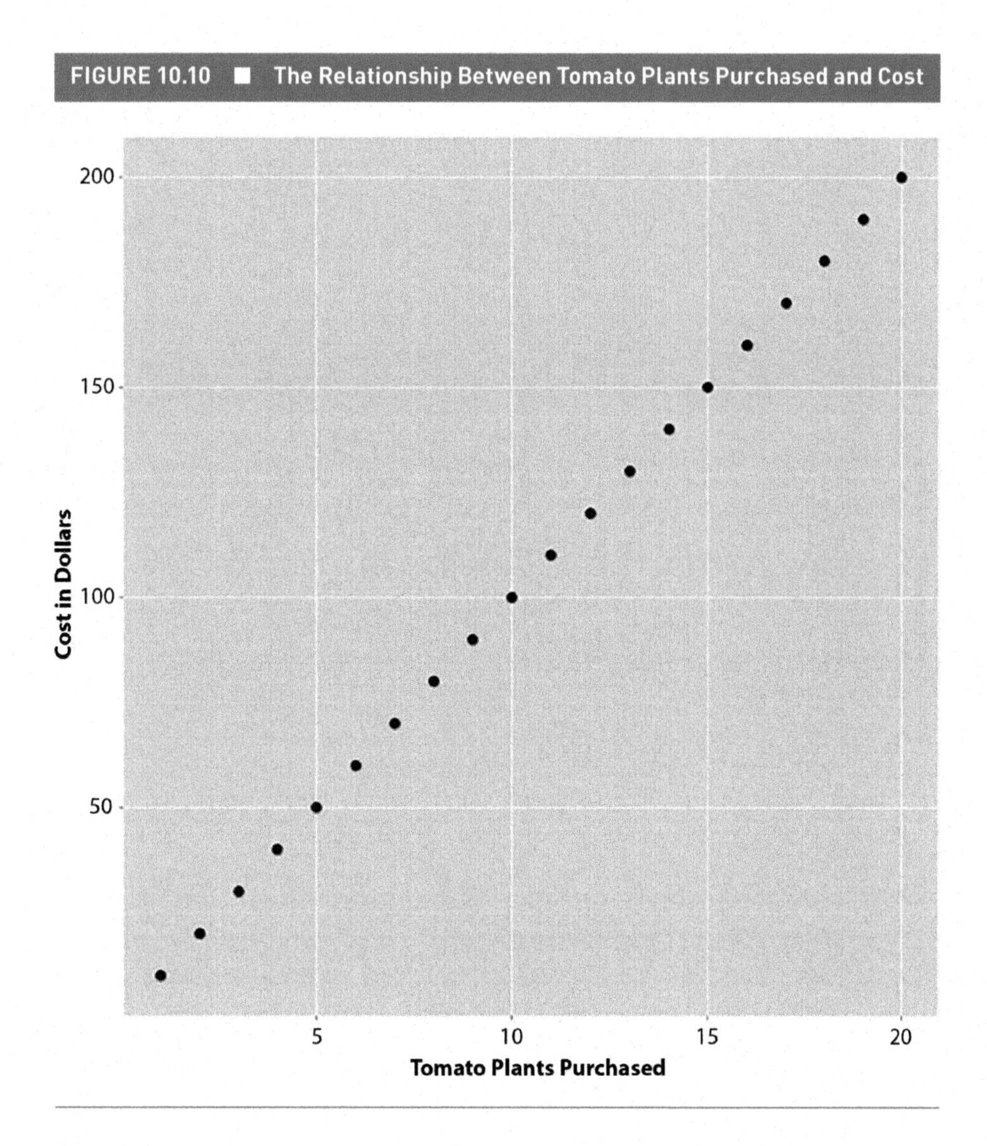

coefficient r can range in value from −1 to 1, indicating the level of linearity between two variables. The closer this coefficient, r, is to −1 or 1, the more likely it is that the relationship under investigation resembles a linear relationship. The closer this coefficient is to 0, the less likely it is that these two variables are correlated to each other. Correlations of 1 or −1 are perfect positive and negative relationships, respectively, while a correlation coefficient of 0 indicates no relationship between variables. By looking at the three graphs in Figures 10.11 to 10.13, we can see how a negative, a positive, and no relationship between variables look when plotted in a graph.

FIGURE 10.11 ■ Positive Relationship Between Father's Height and Son's Height

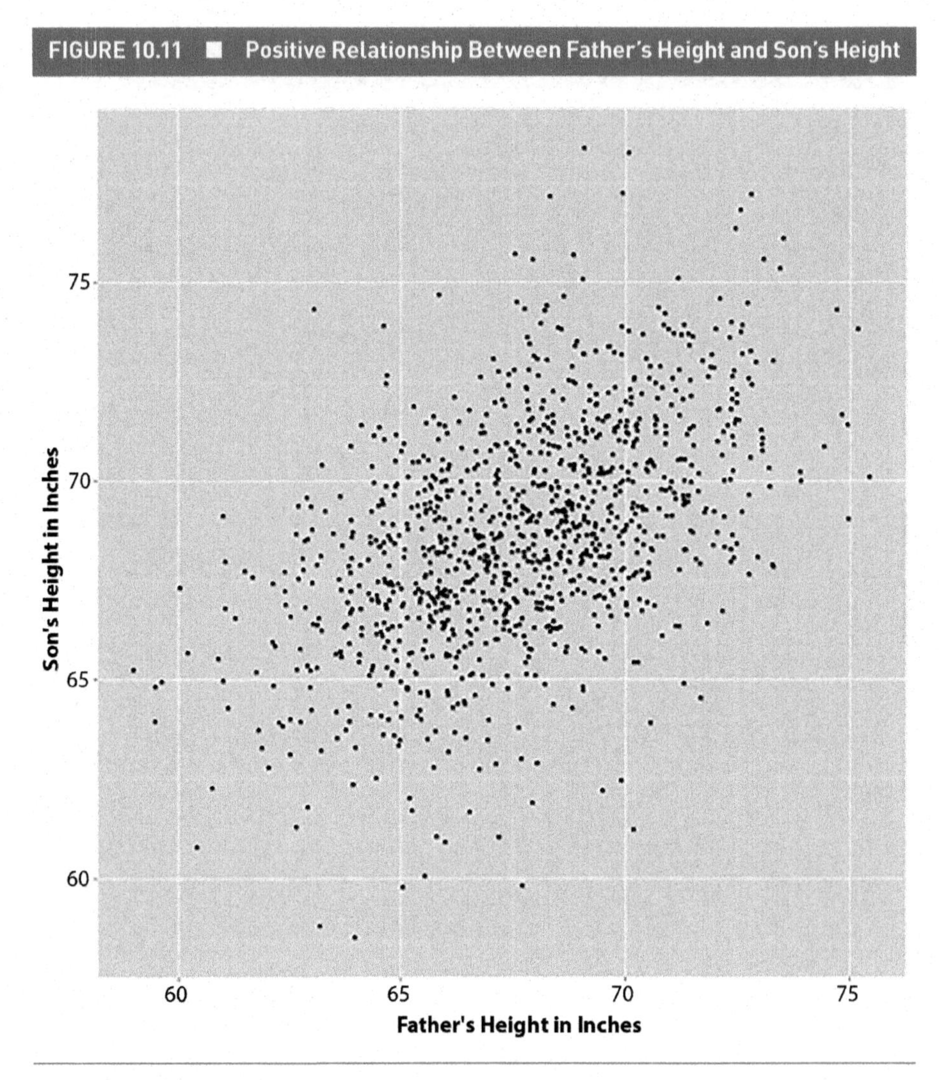

Source: Data from UsingR free packages.

Causation

Finding a strong correlation between two variables in a study can be very gratifying, especially if it was a predicted relationship. Many discoveries in science are based on researchers' ability to prove the relationships they have predicted between variables. Being able to state that eating a plant-based diet is related to better health is based on a correlation between what food participants eat and their overall health. The fact that two variables are strongly correlated with one another does not mean, however, that

FIGURE 10.12 ■ No Relationship Between Mothers' Age and Their Height

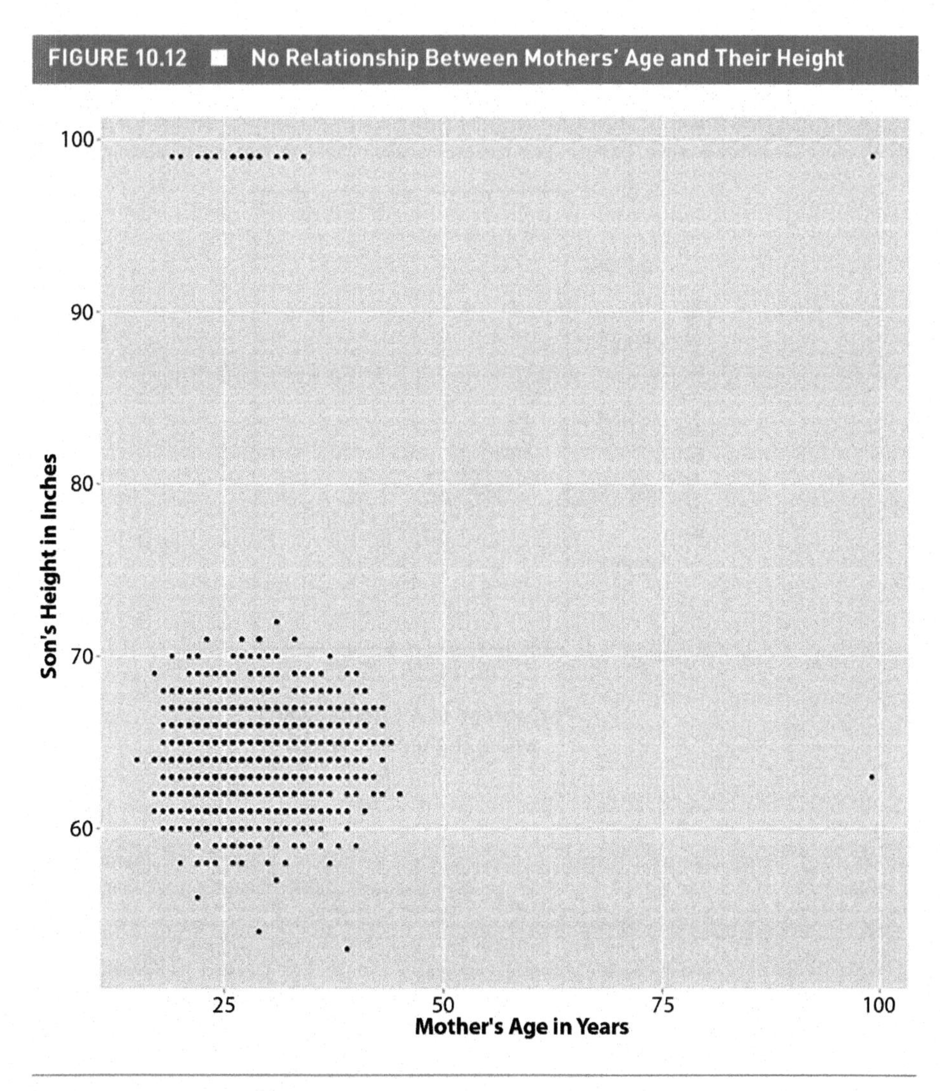

Source: Data from UsingR free packages.

one variable causes changes in the other variable. This is an important point. We cannot conclude, for instance, that the food that people ate contributed to their improved health, or that people who have better health in general are more drawn to plant-based diets. It could also be that people who had better health exercised more, which, in turn, kept them in better health, and the relationship between the food and health was a *spurious* one. A spurious relationship happens when the correlation between two variables is affected by a third variable that interferes with the relationship we are studying.

FIGURE 10.13 A Negative Relationship Between the Percentage of all Boston Homeowners Who Have a Higher Net Worth Than the Homeowners in a Specific Neighborhood and the Price of Houses

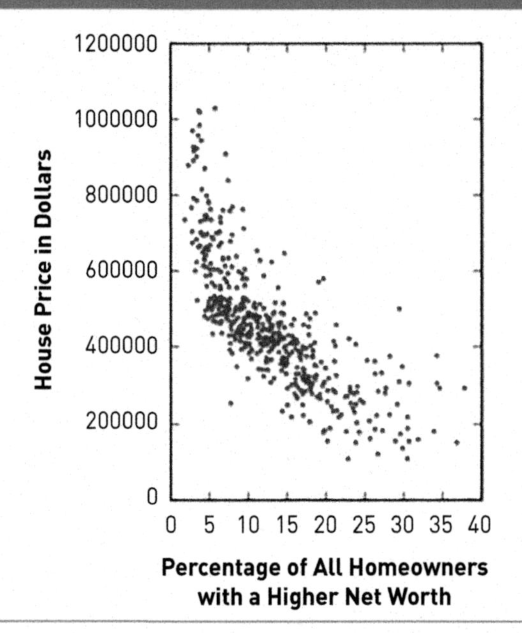

Source: From Boston Project dataset of Machine Learning Repository.

RESEARCH WORKSHOP 10.1

HOW TO FIND CORRELATION USING EXCEL, R, AND SPSS

A correlation analysis is easily run in any software. Once the data have been entered in a spreadsheet, a researcher can run frequency distribution, look at the variables, and run a correlation analysis to see whether two variables are associated with each other.

In Excel, a correlation can be run by using the following steps:

1. Click on the Data tab and then on the Data Analysis tab.
2. Click on Correlation.
3. Select the two variables you would like to see an association for and hit Enter. Another way to do this is to click on the Function button. Under Statistics, you will find Correlation, and then follow the steps 1 and 2.

In R, you first import the dataset from an Excel file or another file you have saved. If you are working from R-studio (also free), to find the correlation coefficient, you need to write the following formula:

cor (datx, daty, methods = c("pearson")) or simply cor (datx, daty)

Because Pearson correlation is the default formula. In this formula, cor stands for correlation; dat$x means the variable x is in the dataset called "dat." These names can be anything, such as parental$time where "parental" is the name of the dataset and "time" is the name of the variable inside the dataset. The same rule applies for other variables of interest.

If you are using SPSS and have your dataset uploaded, go to the Analyze tab, and click on Correlation and then Bivariate. Here you will select the variables you want to test for correlation. Once you have selected the variables, you will also choose the type of correlation—in this case, "Pearson." Before clicking OK, you may click on the Options tab, where you may find means and standard deviations by simply selecting the option. Then return to the previous box and click OK. An output will pop up with a table that contains all the information requested.

Regression

A perfect linear relationship almost never exists, so our data will rarely create a straight line when plotted in a graph. However, we can draw an imaginary line that will best reflect the data points. This imaginary line is called the **regression line** or **the best fitting line** and represents the closest possible linear relationship between two variables. Conceptually, regression shows one line that could have best explained the relationship between variables if this were a linear relationship.

The regression line is expressed by the formula $Y = a + bX$, where *a* represents the intercept of the y-axis and *b* represents the slope of the relationship. The intercept, expressed by *a* in the formula, represents the point where the line cuts through the ordinate or the y-axis. The criterion for finding the values of the intercept and slope is formally known as *least squares criterion* because the regression line represents the one line that can connect the smallest distance of each squared data point. It is an approximate line that attempts to capture the trend of the data.

Regression is used to predict future behaviors of the variables as much as to explain current behaviors. The best fitting line of the relationship between variables also tells us the direction of a relationship. For example, the line of best fit can imply that if we follow a plant-based diet, we are more likely to have better health.

Regression and correlation in data analyses and prediction can be extremely useful for improving our society, from minor circumstances to those that have a high impact on our lives. For instance, a high correlation between body posture and smiling and the chances of being hired in a job interview may determine our body posture and our frequency of smiling. Similarly, a negative regression between the number of hours of television watching and toddlers' ability to speak and interact with others suggests that children should not watch television until they are at least 2 years old.

It is a great benefit to accurately predict a relationship between variables when linear, but what about when the relationships are not linear? How can we better understand relationships between variables that, if plotted, look like a U-shape, are curvilinear, or look like loops? Pearson correlation is only useful for linear relationships, but we are still able to assess nonlinear relationships between variables. Statisticians and data scientists use a polynomial regression or curvilinear regression to determine the relationship between variables with an obvious pattern that cannot be expressed linearly. It is beyond the scope of this textbook to explore curvilinear relationships, but if you are interested in understanding more about the topic, refer to free courses from udacity.com on descriptive and inferential statistics.

Scatterplots

Scatterplots visually depict a bivariate relationship. Scatterplots are dots on a graph that represent a relationship between two quantitative variables: dependent and independent. Each single point represents one case; we usually put the dependent variable on the vertical axis and the independent variable on the horizontal axis. By looking at all these cases as points on a graph, we can understand the relationship between two variables.

For example, the scatterplot in Figure 10.14 shows the relationship between body fat and weight in 252 male participants (taken from UsingR free educational datasets). The researcher explores whether weight relates to an increase in body fat. The body fat variable is measured using Brozek's equation, whereas weight is measured in pounds. All the points demonstrate the relationship between weight and body fat. Since all the points congregate in one corner of the graph, we understand that a higher weight indicates a higher body fat for these male participants.

Outliers

Outliers, as we previously discussed, are the extreme values in a dataset. They are always present, including in our data. When measuring shoe sizes, we know that considering age and gender, people's feet are going to fall within a specific range. However, we also know that there will be people who either will have much smaller or much larger feet than their age or gender group, will need custom-made shoes, or will have to shop in a different department of the store than the others in their age or gender group. These cases are known as outliers.

FIGURE 10.14 ■ Scatterplot of Body Fat and Weight

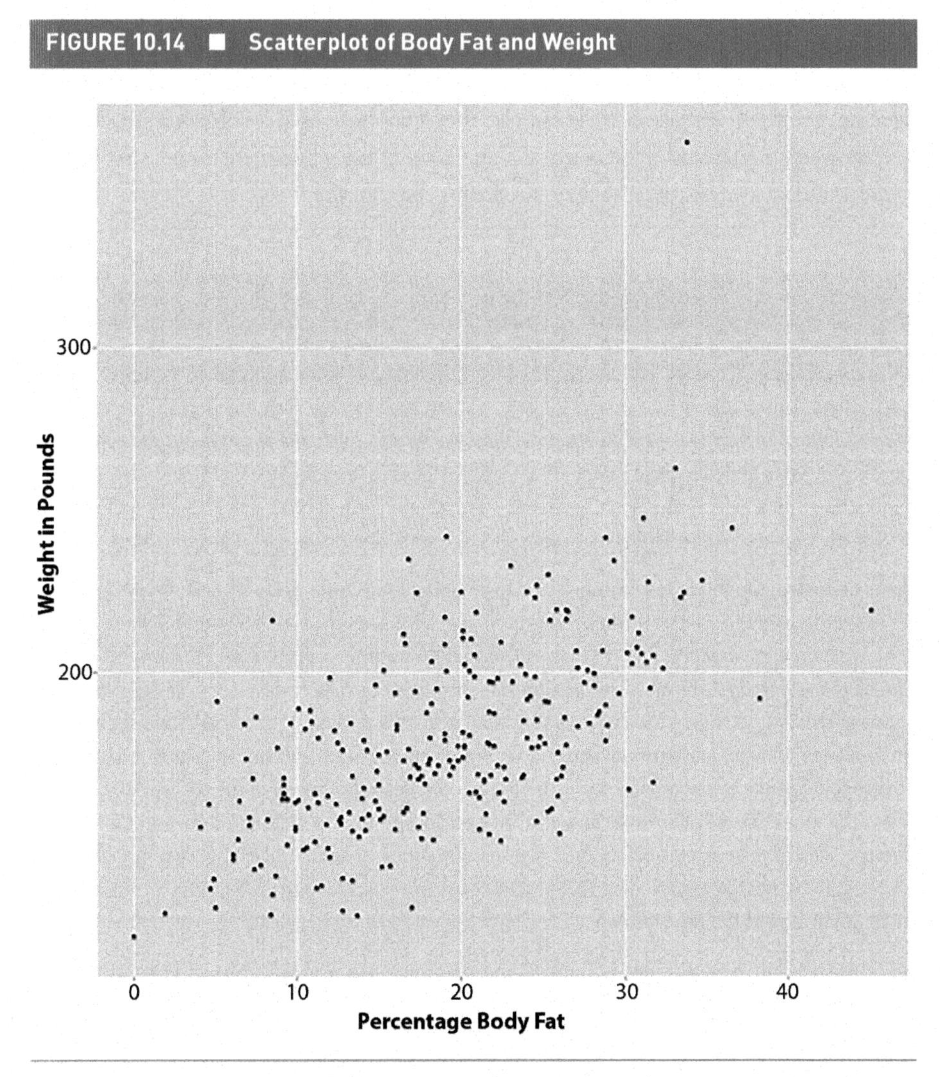

Source: From the dataset FAT from UsingR free educational datasets.

When measuring the Pearson correlation for a dataset, we should follow the steps shown for finding the IQR to detect the presence of outliers. Let us consider a relationship between socioeconomic status (SES) and alcohol use. Elliott and Lowman (2015) conducted a study that investigated this relationship among 4,979 participants. They discovered that a higher SES (higher income and higher education) was related to lower alcohol consumption, but also noted that the religiosity that was more prevalent among lower SES participants served as a variable against alcoholism. If the researchers had a subset of people with high income and education who also drank a lot of alcohol, their findings could have been distorted, even though this small subset of people would not have been representative of the entire population of high-SES participants. If Elliott and Lowman were not aware of outliers in their sample, they may have used a correlation coefficient that showed no relationship between SES and alcohol use when there is one. Outliers are a threat to data analysis because they can distort research findings.

Summary

Univariate analysis is the analysis of one variable, whereas bivariate analysis is the analysis of the relationship between two variables. Univariate is often descriptive in nature because we are considering one variable only. Descriptive statistics gives the researcher a good sense of who is participating in the study, which age group, and many other types of descriptive information. It helps to gain familiarity with the dataset. One common way of conducting univariate analysis is a frequency distribution. We can visually present univariate analysis of the variables of interest by using bar graphs, histograms, line graphs, and pie charts.

We use the measures of central tendency to summarize one variable by expressing it with one single value that is the most representative of the variable. Such measures are the mean, median, and mode. The mean is the mathematical average of a variable. The median is the middle number, where 50% of the numbers fall below and 50% of the numbers fall above its position in the variable. The mode is the most widely used number in that variable. We also use measures of variability to make sense of our variables, such as standard deviation, variance, interquartile range, and range. Standard deviation and variance both express the approximate average distance of each point from the mean, whereas the range is the distance of the largest number from the smallest number. Interquartile range represents the distance between Q3 and Q1 and is approximately 50% of the variable. Q3 and Q1 are the midpoints of each half of the variable when ordered from the lowest to highest score. We use IQR to detect the presence of extreme points or outliers in either the lower or upper end of the distribution.

Part of getting familiar with our own dataset is understanding how close our distribution is to a normal distribution. The normal distribution is a theoretical concept that shows a perfect distribution of the scores where the mean, median, and mode fall at the same point and 50% of the points are above and 50% are below. Z-scores identify the number of standard deviations in the normal distribution. In addition to the normal distribution, we can encounter distributions that are skewed on the right or the left in a variable. We can have a positively skewed distribution where the highest frequency is located on the left side. In this case, the mode is the smallest in value, then comes the median, and the largest is the mean. The opposite happens in a negatively skewed (left-skewed) distribution where the highest peak or frequency is located on the right side. In this case, the mean is the smallest in value, followed by the median and finally the mode.

To conduct more meaningful analysis, we use bivariate analyses to determine the relationship between two variables. One of the basic relationships explored in statistics is the linear model or the ability to predict that when one variable changes with one unit, the other variable will also change. We use correlation and regression to understand linear relationships between variables. Correlation is expressed through a Pearson correlation coefficient that tells us how close to a linear relationship two variables have. If the coefficient is close to −1 or 1, we can have a perfect linear relationship. If the coefficient is close to 0, we can say that the variables are not related in a linear way. Having a correlation coefficient of 0 means that the relationship is not linear, but there is always possibility for other types of relationships like curvilinear. Regression expresses the best-fitting line that connects the changes of one variable as expressed into another variable. The regression line is an imaginary line that demonstrates how the data are related.

Key Terms

Bar graph: a graphical representation that uses vertical bars to represent each variable.

Bivariate analysis: used to investigate the relationship between two variables.

Correlation: used to determine the extent to which two quantitative variables approximate a linear relationship.

Cumulative percent: the collective percentage of categories together.

Descriptive analysis: analysis that allows researchers to sketch the details of variables and gain familiarity with the sample.

Frequency distribution: a simple procedure that can illustrate key characteristics of participants in the sample.

Histogram: a graphical representation that presents the frequency of a variable in the form of distinct but connected bars.

Inferential analysis: analysis that provides a deeper understanding of variables because it draws conclusions about the population from the sample.

Line graph: a line connecting points in a distribution.

Mean: mathematical average of a set of numbers.

Measures of central tendency: summarize an entire variable in a single value that is numerically representative of this variable.

Median: middle point in a distribution.

Mode: number that occurs with the highest frequency in a distribution.

Outliers: extreme values in a variable either too high or too low.

Pie chart: a graphical representation of the distribution of one variable, which works best with categorical variables with a small number of categories.

Quartiles: division of a distribution into quarters.

Random error: error that happens randomly during the process of handling data.

Range: the difference between the maximum and minimum values in a set of numbers.

Regression: identifies the line that best describes the relationship.

Regression line: an imaginary line that best reflects the data points.

Sampling errors: errors or mistakes that happen during data collection.

Skewness: asymmetrical distribution, in which the distribution is skewed to the left or right.

Standard deviation: the square root of the variance that represents each data point's average deviation from the mean.

Statistics: a branch of mathematics in which logic is used to interpret phenomena and draw conclusions from accumulated data.

Univariate analysis: the analysis of only one variable; often descriptive in nature.

Variance: a measure of how spread out every point in the variable is from the mean.

Taking a Step Further

1. What type of variables are best for frequency distributions?
2. What type of variables are best for regression analyses?
3. Can you name an example of univariate analysis? How is that different from bivariate analysis?
4. What type of information can we draw from standard deviation? For example, how can you interpret the following information: "The average age of students in our campus is 18 with a standard deviation of 2.5 years."
5. How are correlation and causation different and similar?
6. What does the regression line or the line of best fit tell us about data that are not yet collected?

QUALITATIVE DESIGNS AND DATA COLLECTION

Understanding What Behavior Means in Context

BY CHARLES SARNO

CHAPTER OUTLINE

WHAT WILL YOU LEARN TO DO?

1. Understand the differences between quantitative and qualitative approaches to research
2. Familiarize yourself with participant observation and in-depth interviewing
3. Gain knowledge on other sources of qualitative data
4. Get exposure with unobtrusive methods, including content analysis
5. Understand how research can employ mixed methods and the benefits of doing so

AN INTRODUCTION TO QUALITATIVE RESEARCH DESIGNS

In Chapter 4 we discussed how to design and collect data for a quantitative research project. As we saw, quantitative designs are most useful for describing the frequency of certain attitudes and behaviors in a large population, as well as for measuring the presence of some proposed theoretical relationship between variables. By definition, quantitative research presents results in some numerical form, and typically employs various types of statistical analyses to explain the strength of the relationship between two or more variables. Qualitative research, in contrast, relies heavily on descriptive language in order to convey its central findings. It seeks to represent the distinctive qualities of the social phenomena under consideration, often focusing how members of the group understand their own subjective experience. Qualitative researchers employ ethnographic fieldwork and intensive interviewing as the primary techniques for gathering information.

As an example of this distinction between quantitative and qualitative approaches, consider a recent Pew research poll, which found that 89% of U.S. adults believe "in God or a universal spirit," and that 63% are "absolutely certain" in this belief (Lipka, 2015). This is a prime illustration of how quantitative survey research can numerically describe the characteristics of a large population. Moreover, an enterprising researcher, interested in testing different sociological theories of religion, could come along and possibly use the Pew data to create a quantitative design which examines the effect certain demographic characteristics, such as gender or age, might have on this belief in God. It might be hypothesized, for instance, that women are more likely to believe in God than men, or older people are more likely to believe than younger ones (see Murphy, 2016). As you can see, quantitative designs are well poised to answer these sorts of questions by providing a precise statistical description of the "who, what, and why" of certain social phenomena. Unfortunately, what is missing from this quantitative equation is a deeper and more nuanced understanding of what exactly it is that these survey respondents mean by the term *God* or *universal spirit*. We might suspect at the outset that not all members of the population mean exactly the same thing. For some, God might be a personal being with whom they have an intimate personal relationship, while for others God might be a more remote and impersonal force. Some might believe that God is all good, loving and compassionate, while others might believe God is vindictive and judgmental. Quantitative research, for all its evident strengths, has difficulty resolving these sorts of questions about meaning and providing a deeper understanding of the context within which such meanings emerge and are played out. Answering these types of research questions requires a more a qualitative research design which employs a different set of techniques for collecting data.

Qualitative research, as the name suggests, is more interested in describing in detail the complex qualities that make up social phenomena, and how these qualities relate to one another and have meaning within a broader, more holistic cultural context. Generally, the qualitative researcher is interested in trying to *interpret* and *understand* behavior within its context rather than trying to *explain* it. Qualitative research operates under the premise that human behavior is subjectively meaningful for the people engaged in that behavior, and it attempts to provide a fuller understanding of that behavior. There is an old Native American proverb: "Don't judge another person until you've walked a mile in their moccasins." In many respects, qualitative approaches can be seen as attempts at "moccasin walking" in order to understand the experience from the perspective of the participants involved, from the "inside out" as it were.

According to the famous anthropologist Clifford Geertz (1973), the primary goal of qualitative research is to provide a "thick description" of the attitudes and behaviors under investigation. A **thick description**, in addition to offering a vivid account of the social interactions taking place between people, seeks to put those interactions against the wider cultural framework that serves as the background for further interaction, while showing how things "make sense" from the perspective of the participants involved. For the qualitative researcher, behavior has no intrinsic meaning independent of situation and setting. Consequently, it is necessary to understand the intentions and consciousness of the people interacting within that setting, and how all this creates the framework for further interaction. To illustrate this point Geertz uses his now classic example of "the wink," which I will appropriate here for my own ends.

Imagine during the course of his or her lecture about qualitative methods you see your instructor wink at another student in class. You would no doubt wonder, "What's up?" and "Why is the instructor doing that? What does that wink *mean*?" Perhaps your instructor just has a piece of dust caught in his or her eye, and the wink is simply an involuntary reflex and thus an innocent gesture. If so, then all is right with the world. But perhaps—goodness forbid—the instructor is trying to flirt with the other student, which doesn't seem very professional or appropriate at all given the situation! After all, the culture of modern education generally frowns on teachers having any romantic involvement with their students because of the unequal power dynamics involved. If *that* is what the wink means, then all is *not right* with the world. At this point, the wink takes on a much more negative connotation because of this ethical concern. But perhaps, instead, the instructor and the other student have some sort of inside joke going on, and the wink is conspiratorial in nature. But what's the joke, and why have others been left out?! It could be that the instructor forewarned the other student that this wink would occur as part of a lesson plan, and not to be disturbed by it. In that case, the other student understands the instructor's signal. It's now time to let everybody in on the joke: The wink is simply the instructor's way of demonstrating how the same behavior can mean different things to different people, depending on the social context and other background factors involved. In this instance, the wink is merely a pedagogical example serving to illustrate the point about thick description. And thus, all is right with the world again. So here we see that the simple gesture of a wink can mean any number of things, and we really cannot understand the behavior without probing more deeply into the intentions behind it, and what it means to the various participants involved.

Given this purpose, qualitative research designs proceed in a different fashion from quantitative ones. For one thing, rather than beginning with a specific hypothesis, qualitative research begins with a more open-ended series of research questions. To continue with the Pew research survey provided above, instead of hypothesizing that "women are more likely to believe in God than men," a qualitative researcher might ask instead, "What do people mean when they talk about God? What do people think and feel about God? How did their ideas about God come about and how do these ideas influence how different people behave? Are there any gender difference in the ways men and women understand God and how they act on that understanding?" As the research proceeds, these questions will become even more refined and focused, taking into account other background factors being observed. But as one can see, at the outset qualitative designs are much more flexible and more adaptive to an unfolding interpretative situation. Qualitative designs rely on an iterative process of data collection and analysis. This means that the researcher reflects on the initial data collected and allows those data to guide the further gathering of information. This flexibility is in marked contrast to quantitative approaches. In order to carry out a quantitative design, the variables must be specified and operationalized well ahead of time. At this point the analysis is locked in place. Moreover, while it is possible to "go fishing" in order to see if any statistical patterns can be found in the data you have already collected, quantitative designs are more typically deductive because they will state a hypothesis derived ahead of time from some theoretical model found in the literature. Qualitative designs, in contrast, tend to take a more inductive approach to the data collected, allowing for the development of theoretical insights. Consequently, qualitative approaches often move beyond exploration and thick description toward theory construction. The researcher is continuously gathering and sifting through the data, looking to see what patterns emerge; we discuss a grounded theory approach below. As a final point of distinction, quantitative researchers present their results in numerical form, often employing descriptive statistics and tables, while qualitative researchers present their results in a narrative form, often using the words of the participants themselves to explore emergent patterns and themes, and use richly descriptive language to place those words within a broader interpretative context. While quantitative designs *count*, qualitative designs take phenomena into *account*.

WAYS OF COLLECTING QUALITATIVE DATA

The two most common data collection techniques employed by researchers working within the qualitative tradition are ethnographic fieldwork (carried out through a mode of direct observation called **participant observation**) and **in-depth interviewing**, which uses elicitation techniques to draw information out of the research subject. The following sections will describe these two modes of data collection in some more detail and provide you with some practical suggestions for how to successfully carry out these techniques. Afterward we'll look at unobtrusive methods of collecting data for qualitative research. With these unobtrusive methods, the researcher has no direct contact with his or her research subjects, but instead looks at the traces—in the form of cultural artifacts—that people have left behind.

Ethnographic Fieldwork and Participation Observation

Ethnographic fieldwork has been a methodological staple of cultural anthropology since its inception as an academic discipline toward the end of the 19th century. During its formative years, anthropology was concerned with chronicling the culture of non-Western tribal groups. Because of colonialism and modernization, traditional indigenous groups were under the threat of cultural extinction by Western imperial powers. As an exercise of cultural preservation, anthropologists would spend many months—sometimes years—living within these tribal communities, learning the language and chronicling the customs of the people under threat. The final product of this research was usually an **ethnography**, a piece of cultural writing which provided a thick description of the beliefs, practices, and overall ethos of the group being studied. These ethnographic accounts would endeavor to show how particular cultural patterns made sense for their adherents, and would try to translate the internal cultural logic of the group for the outside observer. In this way, beliefs and practice, which may have at first seemed strange and exotic, were made more familiar and understandable through a process of contextualization and cultural relativism.

Likewise, the discipline of sociology, which took as its object of study the modern industrial societies emerging in Western Europe and the United States, has a rich tradition of fieldwork. In particular, during the first half of the 20th century the Chicago School of sociology became renowned for using participant observation to study urban subcultural groups like youth gangs, ethnic neighborhoods, taxi dance halls, jackrollers, hobos, and other such "exotic" groups (Deegan, 2007). Many of the first Chicago sociologists were initially trained as newspapermen, and they brought many of the same journalistic techniques of keen observational and interviewing skills with them as they gave account of the different subcultures that composed the urban landscape. Many sociologists today continue to use ethnographic fieldwork as their primary mode of data collection.

When conducting fieldwork, a researcher will enter into a particular social setting and spend an extended period of time (at least several months) interacting with members of the group in question in order to learn about their entire way of life. Conducting ethnographic fieldwork is a long, laborious process that details as much information as possible about a population, culture, phenomena, or selected cases. Ethnographers spend considerable time with their subjects, recording information and endeavoring to understand their world view. Simply put, ethnographers attempt to put themselves in the position of their participants in order to comprehend their behaviors, actions, feelings, customs, and other social phenomena. Being in the field for an extended period of time allows the researcher to observe a wide range of behaviors and to probe the meanings behind those behaviors. Moreover, as group members become familiar with the researcher, **rapport**—a sense of trust between the researcher and the subject—tends to develop. Over time group members take for granted the presence of the researcher and are more willing to open up and behave more "naturally" and honestly. This allows for a richer, more complex, and nuanced understanding of the setting to emerge.

During the course of fieldwork, the researcher becomes a participant observer within the group or organization being studied. The specific role one assumes can take several different forms, ranging from (1) a complete participant, to (2) a complete observer, to (3) some combination of participant-observer. A **complete participant** is the researcher who is not simply observing a population or event of interest, but is also experiencing everything that participants are experiencing firsthand. Often the researcher might already be a member of the group being studied. While the complete participant role suggests that one is intimately familiar with the group being studied, there is the danger of what anthropologists call "going native." This means that the researcher is so close to the culture that he or she cannot write about it in an objective, analytic, and dispassionate fashion. A **complete observer,** on the other hand, is the researcher who is simply observing without participating or experiencing the dynamics that participants are going through. While this allows for some analytic distance and a greater degree of objectivity, the researcher may not fully experience the group's culture, nor fully understand and appreciate all that's involved with the experience. Most researchers opt for some combination of these two positions: the **participant-observer.** This combination allows for greater flexibility and balances the goals of both representing the subjective experience of members, but placing that experience within some broader analytic framework. Depending on what is taking place in the setting, researchers sometimes lean more toward the participant side of the continuum, other times more toward the observer side. The exact combination they adopt is usually contingent on the nature of the scene under study, what makes the most sense analytically, and their own personality type (e.g., are they an introvert or an extrovert?), which influences their choice of involvement.

ETHICAL CONSIDERATION 11.1

FIELD WORK

Ethical issues abound during the course of fieldwork. This is because the researcher is engaged in direct, constant, and continuous social interactions with the subjects of research. One issue that often arises during the fieldwork is the exact nature of one's role within the setting. Are you going to be more of a participant or an observer? Is your research role going to be completely overt or might it have a covert (i.e., "undercover") dimension? It is strongly recommended that you get permission to study the group from some gatekeeper or group leader and let members know you are there in the setting as a researcher. But should you preface every informal conversation you have with people by reminding them of this fact? Moreover, while longstanding members of the group are likely aware of your role and intentions, what about newer members? How will you inform them of your role in ways that don't seem awkward or off-putting? Even if you are clear about your role as a researcher, be aware that for some people you may be more than that, becoming a friend and confidant. Others may become the same for you. How then should you respond to them? For example, if you are told something in confidence, does that become part of the data that are being collected? How should you respond to political divisions which arise in the group? Should you take sides? It is likely that if you spend an extended period of time

(Continued)

(Continued)

doing fieldwork in a particular social setting you will find yourself occupying multiple roles. This might occasionally cause confusion and potential misunderstandings between you and some of the members of the group you are studying, because you may have one set of expectations about the role you are playing, while they might have other expectations.

As a further illustration, let me discuss just a few of the ethical predicaments that arose when I was doing fieldwork in the setting of a Christian fundamentalist church. This was a very conservative religious organization whose members confessed Jesus Christ as their Lord and Saviour and believed that their interpretation of the Bible provided the only true pathway to salvation. I personally did not subscribe to this church's belief system. Consequently, I was more of an observer rather than an active participant in the group. Nonetheless, I attended services regularly and engaged with members in prayer and Bible study because, as a researcher, I was interested in maintaining rapport with members in order to achieve a fuller understanding of their faith. While I oftentimes disagreed personally with what people may have believed and some of the things they said, I did not feel it was my place to be contentious or engage in debate. For one thing, this would have adversely affected my rapport with members and undercut the empathetic understanding I was attempting to achieve. But note that for me empathy did not indicate sympathy with the belief system. While some members of the church could appreciate my position in the role of researcher, others were confused and even threatened by it. "How can you understand my faith so well and yet not still not believe it?" some would ask. It was difficult to provide a clear and straightforward response in a way that would not alienate members in this setting, and this presented an ethical quandary: How honestly do I present my own thoughts and opinions on these religious matters?

While it's generally advisable to be as open and honest as possible with your subjects, this sometimes is easier said than done. Much of what is considered an ethical course of action will depend on you, the sometimes competing values you hold, and how these values comport with the values of the people in the setting you are studying. It is strongly recommended that you check in periodically with an instructor or advisor to discuss the ethical dilemmas that may arise during the course of fieldwork, and to frequently use your field journal as a venue for exploring and working through any issues which arise.

In the course of fieldwork some group members will be extremely helpful in "showing you the ropes" of the setting. These members are your **informants.** They can assist you in making connections with other group members, understanding the various beliefs and practices you are observing, and generally making sense of what's happening. During this time, you are keeping a **field journal**, writing down everything you are observing and experiencing in the setting. The field notes which compose this journal can take a variety of forms. First and foremost, they provide detailed descriptions of everything the researcher is seeing and hearing. Field notes should be written as close in time to the actual experience as possible, and no later than 24 hours after the fact, because after that time one's memory will fade significantly and important details will be lost. The initial period spent in the field can feel overwhelming, and field notes are an important tool in processing the experience in writing and putting things into perspective. Because you may not understand what is important when first starting out, include as much detail as possible; it may prove significant at a later date. Field notes are also a place for methodological and theoretical reflection. As time goes on, the notes will become less descriptive and more reflective, a place for thinking about the credibility of the information you are gathering and putting that information within some broader analytic framework.

RESEARCH WORKSHOP 11.1

ENTERING AND EXITING THE FIELD

Fieldwork is a research process that generally unfolds in stages over time. While the bulk of fieldwork takes place during the middle stage, the beginning and end of the process are important to think about. When entering the field there are several things to consider:

1. You'll start with some fieldwork setting in mind. Is this setting a private or a public one? If the setting is private—like a church, a hospital, or a motorcycle gang—**you will need to get the permission of the leader or some authority figure in charge of the group or organization under consideration.** This person (or sometimes persons) in charge is **the gatekeeper** for the organization. Even in the case of public spaces, you'll want to think about any authorities you may need to consult. Perhaps you are interested in observing shopping patterns at the local mall. While this project seems innocuous enough, realize that scoping out public places for an extended period of time may raise concerns (in this age of terrorism) among those in charge of security in the setting. When talking with any gatekeepers, you'll want to be able to give a broad overview of your project so they, in turn, can give you their informed consent to participate.
2. Once you've secured permission to be in a research setting, you'll want to consider **how much to read about the setting ahead of time.** Part of this decision will depend on whether you plan to use the grounded theory approach to research, discussed below, or the extended case method. If you are using a grounded theory method you should not read too much ahead of time so as to avoid any preconceptions about the setting. With the extended case method, performing a prior literature review will be necessary so that you will be theoretically and conceptually sensitive as to what you are looking for during your research.
3. Finally, **how will you know when it's time to leave the field?** This usually occurs when you reach a point of **saturation** and **are no longer seeing anything new or unexpected** during the

(Continued)

(Continued)

course of your research. Just like a sponge that can no longer sop up any more liquid, you as the researcher are not able to absorb any fresh information. When you're at this point, you can pretty much predict the routine events and activities that occur in your setting and the typical responses of the participants.

4. You will likely have formed close relationships and become friends with some of your informants, and **you will need to decide in what way, if any, you wish to continue those relationships**. Also, it's not uncommon to think about how you might give something back to the group whose members have shared their time and lives with you. It might be by providing some assistance to the group or as simple as sharing a copy of the ethnography you produced.

Ethnographic fieldwork is usually invested in exploring and understanding the details of how something happens or how something is experienced rather than testing a theory. Nonetheless, while the purpose is often descriptive, ethnographies can bring forth new theories or understandings of social phenomena. Theory development is usually done inductively by comparing and contrasting patterns that you think you see emerging in the data. This inductive approach is called the **grounded theory method**, as more conceptually abstract ideas about what is happening are developed from the ground up, from the detailed observations that are constantly being sifted and sorted through in one's field journal during the course of research.

This grounded theory approach to qualitative data, pioneered by Glasser and Strauss (1967) is discussed more fully in Chapter 12 in relation to analyzing in-depth interviews.

There is an alternative technique to theory development in fieldwork known as the **extended case method**. This method, advocated by Michael Burawoy (1998), employs a more deductive approach as it looks at other instances or cases of the phenomena under investigation to see what existing patterns have already been found and the theory used to explain them. The researcher then uses his or her own field site as a test case to see what the limitations of the existing theory are and how it might be further refined and extended so as to apply to his or her own setting. The word *case* here may refer to participants, particular events, groups, cultures, and other phenomena. If different patterns are found between the researcher's observations and those in the other case studies, the researcher would then try to discover the other missing factors involved that would help better understand the behavior. In this way, the extended case method attempts to avoid having to constantly go back and reinvent the theoretical wheel during fieldwork by building on existing insights and models found in the literature. This method does presume that there are already existing ethnographic accounts of settings similar to the one the researcher has identified. Often this is not the case, which is why many qualitative researchers tend to use the more inductive approach of grounded theory.

In-Depth Interviewing

In addition to the participation observation method employed in ethnographic fieldwork, another common way of gathering qualitative data is through in-depth interviewing. The **in-depth qualitative interview** asks participants a series of open-ended questions which probe a topic with some focus and in some detail. Typically, in-depth qualitative interviews are recorded and then transcribed later. The analysis of such data is described in Chapter 12. In-depth qualitative interviews may be **structured, semi-structured,** or even **unstructured**. The differences depend on how detailed and worked out the **interview schedule** is ahead of time and how strictly the interviewer follows the schedule. Simply, the interview schedule is the series of questions asked of the interview subject.

In the case of structured interviews, the questions composing the interview schedule are fully worked out ahead of time, and the researcher adheres strictly to the protocol without any deviation. The idea is that all participants are asked the same questions in exactly the same way and in the same order. Although participants may give diverse answers to the questions at hand, and often researchers prepare follow-up questions to possible answers, all of the questions are formulated and presented in the same way to every participant.

On the other end of the continuum, the researchers conducting unstructured interviews may not have an actual interview schedule at hand, but rather just a few broad topics and questions in mind that they plan to cover. Unstructured in-depth interviews are similar to a leisurely conversation between the participant and the researcher. The researchers usually have a clear idea about the focus of the study and what it is important to find out, so they may guide the interview toward their interest, but they also are free to explore other interesting paths on the way. Often the researchers may be intrigued by some new direction that the conversation is taking. In unstructured designs, researchers follow up what they find worth exploring further. While this provides maximum flexibility, it's likely that not all interview subjects will be asked the exact same series of questions in the same way, and thus some standardization is lost in the process.

The semi-structured interview represents a "middle way" of sorts: The interviewer uses a well-worked-out schedule which serves as a clear guide, but then has the freedom to adapt during the course of the interview by skipping or adding questions as the situation warrants.

Best interviewing practices should feel like a leisurely conversation between the interviewer and the interviewee.

Semi-structured interviews have some basic questions that will be asked to every single respondent in the study. Sometimes the researcher may **probe** the participant to say more about a specific issue, while at other times a follow-up question may not be necessary. Semi-structured interviews are extremely useful for gathering the same information from all the participants. Nonetheless during an interview, the discussion may digress to different topics and take the interviewer to unexpected places. This approach is most recommended for the novice researcher, as it provides enough structure to maintain a clear focus and conduct a successful interview without being overly rigid.

When constructing your interview schedule, you should use open-ended questions rather than yes/no questions as much as possible. If questions are answered in the yes/no form, follow up with "how" or "why" so the respondent can elaborate. Similar to when asking survey questions, you'll also want to be clear in what you are asking, and make sure you pay attention to the answers, so you can follow up as necessary. Successful interviewing requires being quick on your feet—a skill that requires practice to make perfect. Finally, avoid questions that may lead the participant in a particular direction. For example, saying "Do you think psychiatric medications harm children's brains?" is not only a bad question because it is asking for a yes/no answer, but it is also a leading question. It is implying these medications are harmful, bringing a judgment into the conversation, and ignoring the feelings, emotions, or situation of the participant. These types of questions can trigger animosity, cause a participant to change their opinion to reflect your own, or cause a shutdown. Another way to put this question would be "What do you think of the types of treatment used for children with emotional and behavioral problems?" If the participants bring up psychiatric medications, follow up by asking, "What do you think of psychiatric medication use for children?"

In all cases, no matter the structure involved, the ability to avoid judgment or bias is one of the most important skills in conducting face-to-face interviews. Just as in ethnographic fieldwork, maintaining good rapport is central to managing a good interview. In fact, both ethnographic fieldwork and interviewing draw on the same set of skills, which again can be learned over time with practice: the ability to be flexible and adapt quickly to changing circumstances, a willingness to pay close attention and really listen to what someone is saying, and a high degree of self-reflexivity which allows one to pull back in order to see how well rapport is being maintained and where the overall interaction is going. Keep in mind that the interviewer needs to obtain necessary information from the respondent, while remaining neutral and not voicing opinions. This includes facial expressions, body language, or presentation and appearance. Sometimes a subtle frown or a smile at an inopportune moment can interfere with an accurate response. Or an interviewer in a suit and tie may project a formal or authoritative tone that may negatively impact certain populations. Sensitive topics can make it particularly difficult to find participants or to approach them. These are groups that may be either involved in or victims of something illegal and are quite vulnerable if their identity is disclosed. Undocumented immigrants, for example, may not be keen to answer questions from someone who is perceived as *authority*.

To reiterate, during the course of the interview it is important to listen carefully and pay attention to what your respondent is saying. While this may seem like common sense, many times inexperienced or first time researchers will be looking ahead on their interview schedule and thus miss important lines of inquiry and follow-up questions because they are not fully engaged with the interview participant. Allow the participant to control the conversation. The researcher can direct the flow of conversation without leading it. The participant is encouraged to show, describe, and summarize his or her opinion rather than the other way around. Your role is to ask questions that will bring out the fullest and most honest possible answers. It is also a good practice to occasionally restate what the respondent told you just to make sure you have understood correctly.

One more tactic to remember about face-to-face interviewing is tolerance for uncomfortable silences. Often a long pause or silence in an interview triggers the urge to speak. The interviewer needs to feel comfortable with these silences because they are likely to encourage the interviewee to break the silence. Often the interviewer may expedite the process prompts, for example, with statements like "Tell me more . . .", or "Can you elaborate more about . . . ," or simply "Hmm . . . ." followed by a long pause.

Although qualitative in-depth interviews are a wonderful tool for collecting data rich in detail, they are prone to biases, sometimes very subtle ones. For example, a researcher may be interested in capturing college students' perceptions of sexual assault, thinking that males will have a different view of the topic compared to females. Imagine a situation in which the researcher had prepared all the questions and follow-ups for the interviews beforehand and was now ready to start the study. But without even being aware of this, the researcher was using a different tone of voice and body language when interviewing women compared to men. This type of problem is common in qualitative work. As the researcher, understand that you are the data collection instrument and need to be continuously monitoring yourself to make sure you are being a neutral and hence more reliable observer. It is something that can be addressed by rehearsing beforehand and being critical and self-aware as to how we interview people. In addition to the suggestions provided here, study the research tips for additional advice on conducting a successful interview.

RESEARCH WORKSHOP 11.2

SOME GOOD PRACTICES FOR CONDUCTING AN IN-DEPTH QUALITATIVE INTERVIEW

Practice is key to conducting a fine interview. One of the instruments you use to conduct your interview—in addition to any recording device—is your interview schedule. Thus, it's important that you know it backward and forward. Run through a practice interview with a fellow student or friend so that you can get a sense of the flow of the schedule, whether your questions are clearly asked and follow a logical order. As you construct your interview schedule, arrange questions in ways that make either chronological or thematic sense. It works best to begin by asking the respondent for some basic demographic information (e.g., age, occupation, highest level of education) before moving on to substantive matters. Ask "easier" questions first in order to build trust before moving onto any sensitive topics. In addition, here are a few other good practices to keep in mind:

1. If you are going to be conducting a lot of qualitative interviews it is worth investing in a decent audio recorder. There are many relatively inexpensive high-quality digital recording devices on the market today. If cost remains an issue you can check to see if your library has recording devices available for loan. **Read the instructions about the features of your equipment and become familiar with how to operate it beforehand.** It's advisable to have the auto time stamp function on the device set accurately, as this will allow you to organize interviews by date later on if you wish.
2. It's always a good idea to **test your recording device ahead of time** to make sure everything is in proper working order. Do you have a fresh set of batteries if needed? Is a convenient power source handy where you are going to conduct your interview? Do you have enough capacity in the device's memory to carry out a lengthy interview?
3. **Consider where you will be conducting your interview**. Typically, you'll want to find a place that is convenient and comfortable for both you and your interviewee. Select a space that is quiet and where the interview will not be interrupted. Make sure the place you select has good acoustics. While coffee shops and restaurants may be comfortable settings, the ambient noises in those public places can make conversations difficult to hear and will also make for

a poor sound quality on your recording. This will make transcribing the interview much more difficult later on. Public libraries often have study rooms available by reservation; these are quiet and comfortable spaces where an interview may be conducted.

4. **At the start of the interview identify yourself and provide the date, time, and location of the interview.** In addition to any written informed consent form you have obtained, it is a good practice to have your respondent provide their full name and again provide consent at the start of the interview.
5. **Keep revisiting and revising your interview schedule during the course of your research** in order to add, subtract, or rearrange questions as needed. One of the general strengths of qualitative research designs is their flexibility, so take advantage of this by reviewing the process from time to time.
6. If you have some time and privacy immediately following the interview, **make some notes as to how it went**: How was the rapport? Were there any especially interesting sections to pay attention to during data analysis? Were there any particular problems? How can you improve things for the next interview?
7. **Once your interview is completed download your audio file immediately,** and maintain a backup file somewhere separate from your download. Give a quick listen to various parts throughout the recording to make sure the sound quality is good.
8. **Be sure to follow up with a quick note of thanks to your interviewee.** Remember, your project is not their project, and you should express some gratitude for their willingness to share their time and experiences with you.

Once you have completed your interviews, you will likely be transcribing them. More will be said about this process in the next chapter, but for now realize that transcription is a laborious time-consuming process. Depending on how quickly you can type, and the quality of your playback equipment, a 1-hour interview can take up to 8 hours to transcribe. So plan accordingly!

Focus Groups

Focus groups are very attractive for researchers because they are cost- and time-effective and yet still include some in-depth insights on participants of interest. The idea of a focus group is basically getting a small group of participants together and interviewing them in a discussion format. The researcher conducting the focus group is called the facilitator and oversees directing the discussion toward the topic of interest. Focus groups are extremely beneficial

Focus groups are quite effective when people who are going through the same experiences are able to share their viewpoints together.

for gaining qualitative insights into a problem in a much shorter time frame, and they work very well for different types of studies. A focus group is very helpful in many instances, such as when the topic of study is controversial and the researcher wants to hear from both opposing sides of a discussion. Moreover, focus groups can also work very well for sensitive topics when participants may not be willing to share their experiences one-to-one, but they may be more inclined to share those experiences if other people are sharing as well. For example, a researcher may be interested in investigating why elderly women stay in abusive relationships and their reasons for hiding the abuse. Now that is a sensitive topic and elderly women who are hiding the abuse to begin with may not be inclined to share their experiences with the researcher What happens if 10 of these women are brought together for a focus group and once the facilitator starts the discussion, only one of them starts talking about the abuse? The other nine participants may be reluctant. However, when one person is describing experiences similar to their own, humans are likely to sympathize and start talking about their own experiences. Therefore, the researcher can now collect information from all 10 women instead of the one or two who were initially willing to talk.

Focus groups require some assistance to arrange and are best set up through the help of a few people. Researchers often use more than one recorder at different points in the room to record the interviews. While focus groups are more efficient in eliciting information compared to one-on-one interviews, they can also be more challenging to manage because multiple participants are involved. In addition, we need to be cautious about specific circumstances when focus groups are not the best option for conducting research. Some of these circumstances include research studies where (1) the researcher is focused on exploring people's behaviors clearly, he or she may want to observe behaviors in specific situations rather than in a focus group with multiple participants; (2) the researcher may have grounds to believe that if people get together in the same room, they may influence each other's opinions and refrain from accurately narrating their experiences; or (3) the researcher's goal is to generalize from the focus group to the larger population, yet the participants may be too similar in their characteristics, and therefore not representative of the entire population.

RESEARCH WORKSHOP 11.3

PREPARING FOR FOCUS GROUPS

Conducting professional focus groups requires time, practice, and specific skills. Here are a few things to keep in mind:

1. You are the facilitator, so **establishing a good rapport and tone for the group without leading or interfering is crucial.** You may need to gently bring out questions, while keeping a low profile.
2. Try to listen carefully to participants and allow other participants to answer or comment rather than answering yourself. **Make sure everyone in the group has the opportunity to participate.** This may require managing the extroverts in the group and encouraging the introverts to speak up.
3. **Avoid biased expressions,** such as "I cannot believe that happened," "This is outrageous," or "How wonderful." Statements of bias might lead the conversation in one direction or another. Although these examples are extreme, even nodding your head may inappropriately lead the conversation.
4. **Make sure that the participants are representative of the population**. Participants should represent a variety of ages, race, ethnicities, backgrounds, and genders. If possible, look for a diversity of thought and opinion, since similarities among group members are problematic for your study, depending on its subject.

ETHICAL CONSIDERATION 11.2

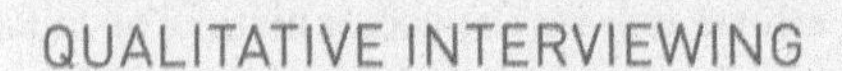

QUALITATIVE INTERVIEWING

Before undertaking any qualitative interviewing, we need to once again revisit the code of ethics for conducting research. Providing interview participants with a written confidentiality statement is one way of informing them about your study in writing. Make sure they are aware that their participation is voluntary and they can withdraw from the study at any time. The confidentiality statement is a simple narrative presented to the participants before data collection. Participants are informed about what the focus of the study is, how the data will be handled by the researcher, and the steps the researcher will take to protect the participants' confidentiality. Typically, in the course of writing up any results the

(Continued)

(Continued)

researcher provides pseudonyms (i.e., fake names) for respondents and removes any information that might personally identify them from their responses. On this confidentiality form, participants are also provided with the researcher's contact information in case they have an interest in reading about the study findings or being further informed about how their personal information was handled. It's best to provide this statement ahead of time to your interview subjects. This helps prepare them for what to expect (e.g., the types of questions you'll be asking, the length of the interview) and helps them feel comfortable with the process. They may be as nervous responding as you are in conducting the interview! Be ready to answer any questions they may have about the process.

Finally, it is possible to mix qualitative methods together. These are simply techniques of data collection, so they are selected to best approach and answer the research question at hand. Many researchers use various combinations of qualitative data collection to best fit the goals of the study. For example, an ethnographic study may combine participant observation with some in-depth interviews. In addition, as will be seen shortly, if members of the group under study have produced any written texts or other cultural artifacts, those elements may also be incorporated into the analysis.

UNOBTRUSIVE METHODS AND ADDITIONAL SOURCES OF QUALITATIVE DATA

The preceding discussion assumes that the researcher has gone out and collected his or her own data through either ethnographic fieldwork or some form of in-depth interviewing. This implies that the researcher will have had some interaction with his or her research subjects. The end result of all these efforts will be the accumulation of much written text produced by the researcher, in the form of either field notes or interview transcriptions. Eventually the results of this material will be presented as an ethnography or reported in some other narrative form. But note that in addition to this textual material produced by the researcher, there are other sources of qualitative data available for research, and these sources of data may be collected unobtrusively.

Unobtrusive data collection means that the researcher has no direct contact with any research subjects during the course of investigation. Instead he or she systematically gathers and examines the cultural artifacts or **traces** that a population has left behind. These traces of human behavior and human thought are an indirect form of observation and can be found everywhere from household items to public information, including media or online sources. Bernard and Ryan (2010) discuss the five common forms of qualitative data which can be gathered and examined in this unobtrusive manner: (1) physical objects, (2) still images, (3) sounds, (4) moving images: videos, and (5) texts. I added a sixth method called content analysis.

The benefit of this unobtrusive approach is that there is no "reactivity" between the researcher and subject that might alter the phenomena under consideration. Reactivity occurs when the participants in a project alter their attitudes or behavior because they are aware that they are part of a study. This is called the **Hawthorne effect**, named after a famous experiment which demonstrated how research subjects often unconsciously change their behavior as the result of being studied. But when a researcher gathers data using unobtrusive methods, the cultural artifacts left behind in this instance were generally produced without any awareness they would be subject to later investigation. While you should be aware that there are some exceptions to this general rule in terms of historical documents and archival data—instances when people may craft written materials to make themselves or their cause look good in the future—in most cases, for most of the qualitative data sources described here, this rule holds. In addition to the benefit of non-reactivity, unobtrusive methods are ideal if you are not a "social butterfly" and don't wish to necessarily interact with other people, but still want to do social research! Using unobtrusive measures allows you to still explore another group's culture by indirectly observing the traces of what they have left behind. In addition, because this type of data generally accumulates over time, it makes it possible to study long-term historical trends. Let's examine in a little more detail some of the types of qualitative data traces that Bernard and Ryan (2010) reference, and some types of studies that might be performed using them.

Physical Objects

Physical objects can be anything from antique pieces, to archaeological ruins, to contemporary household objects that are of interest to the researcher. Some disciplines may be more driven to use physical objects in their studies, such as archaeologists, anthropologists, theologians, and historians. Physical objects are an important source of human development. For example, think of a piece of clothing, furniture, or any other object that was saved for generations in your family that perhaps came from the childhood years of your grandparents or even great-grandparents. If such an object comes to mind, can you describe it in detail? What are your thoughts about it? Can you come to certain analytical and interpretative insights? What does it say about the people who used those objects?

I vividly remember the arrangement of my grandparents' foyer, the two flights of marble stairs leading to their bedroom, the glass French doors with hand-embroidered curtains entering the bedroom, the long mirror right above the shoe drawer, the entire shoeshine kit, the double-sided thick brush that carried other hairbrushes, and the little cup of kerosene on the bureau. All these daily objects are more significant than just reflecting on my grandmother's personal character and cleanliness. They reflect a broader culture—a constellation of beliefs, practices, values, and norms held by members of a group—often expressed in symbolic terms, even though embodied in material objects. In this case, the entire arrangement hints toward the importance of luxury and details on appearance that people held dear as part of their standards. In this cultural landscape, it was also an indicator of the belief that kerosene was good for your hair and scalp, and kept children's hair free of lice. Now, can you imagine how exciting it could be for an archaeologist or historian to have a sample of many such seemingly insignificant objects—foyer gadgets and furniture from those years? With such material objects they could trace the cultural development of a group of people in very subtle and nuanced ways.

Still Images

Photographs, paintings, drawings, x-ray records, or any other kind of images can be an important source of qualitative data. Again, we may see specific academic disciplines that are more likely to use images in their research, but it truly depends on the topic. For instance, communications or media researchers may be interested in looking at pictures of alcohol advertising in newspapers over a 100-year period. A sociologist interested in gender roles may want to see how photos of women in popular magazines have changed from the mid-20th century to the early 21st century. A political scientist may be interested in the changing imagery used by political campaigns during election season, while a medical researcher may want to collect x-rays of people's hands to better understand the onset of arthritis. Images of all kinds are full of details and rich information which allow the researcher to look into the past and see how processes unfolded over time. Such projects are relatively easy to do given the availability of many images and can yield quite interesting results.

Sounds

Audio data, such as voices, utterances, wind, thunderstorms, or any other form of sound is a type of qualitative data. Audio data are widely used by researchers from different fields. Researchers in linguistics may be interested in the development of languages of dialects, while scientists in climate research may investigate the various wind sounds hurricanes generate. Other scholars may be interested in music, tracing the development of different musical genres over time. As has already been mentioned, social scientists often record the data they are collecting via digital voice recorders. In general, these recordings are transcribed and turned into text that is later analyzed by the researcher, but sometimes the actual audio data are explored in their own right if certain tones, various pitches of language, hesitations, or pauses are important to the study. For instance, researchers in communications studies may undertake a "conversation analysis" to understand the different inflections used by men and women in the course of their speech, and they may employ sophisticated audio equipment to measure these differences.

Moving Images: Video

Video data are a combination of audio and images at the same time. This type of qualitative data is widely used in studies where the image and the sound are both important for the investigation. Film researchers or social scientists may be interested in the representation of certain groups or populations over the years through popular movies, news, television advertisements, or any other type of video data. For instance, what percentages of characters in TV sitcoms are members of minority ethnic groups, and do the actors representing these groups have significant parts in shows? Because of smartphones and other video recording technologies, videos of everyday occurrences are now a ubiquitous feature of social life that no doubt will become important data sources going forward. For instance, one could imagine criminologists wanting to analyze the footage taken from police body cameras in order to better understand policing across different communities.

Texts

Written text is the most common type of qualitative data used by researchers in their studies. Although most procedures on how to analyze qualitative data are applied to all types of data, analyzing the written text is the most widely used. Textual material is easier to collect, is cost-effective, and most of the information around us comes in the form of text. We've already discussed the texts generated through qualitative research, such as field notes and interview transcriptions (where audio is turned into textual form). But some texts are created by the subjects of research for their own purposes. For example, letters, journals, and diaries are some of the main documentary sources used by historians. Or consider newspaper accounts, which are often considered "the first draft of history." While the subjects of research might no longer be alive, these textual traces allow us to explore what people thought and felt in the past.

In addition, Bernard and Ryan (2010) identify a few other types of indirect observations: (1) traces of human behavior and thought just described, (2) archival data, and (3) secondary data. Archival data, although similar to traces of human behavior, are a distinct category because they are systematically collected by an institution. In this category, we can find anything from governmental records, such as court documents, minutes of town hall meetings, private information like the cost of patient care collected from hospitals, and cemetery registers. These documentary records can produce a rich history about the institutions that collected them, as well as the particular members who composed those institutions. Secondary data are another unobtrusive method which refers to the systematic collection of information by another party prior to our study. In this category, we can find data collected from the government, national institutes, and other researchers. The Pew Research data cited at the start of this chapter is one such example. One important difference between secondary data and the other types discussed is that secondary data are generally collected systematically and scientifically following a specific research design by the first investigator.

Content Analysis

One particular unobtrusive method that is commonly used and is a wonderful technique for the novice researcher to try out is called **content analysis.** A content analysis is the systematic examination of the trace communications people have left behind, typically in the form of words or images. As we'll see shortly, a content analysis can be done using a mixed-methods approach, and if one can effectively design a mixed-methods content analysis then that person will have addressed all the major research method issues, including sampling, units and levels of analysis, operationalization, reliability, validity, and finally data analysis. For now, I'll only discuss a qualitative approach to content analysis, but below will show how it can be employed using mixed methods.

One example of a qualitative research question that a content analysis could answer is "What are the most common themes found in country music lyrics?" The first step in conducting any content analysis is to define the universe of content available for examination. In this case, I suspect there have been hundreds of thousands of country music songs written over the years. It would be impossible to look at every single one of them. Thus, I would need to narrow the range of possible observations through some method of sampling. Perhaps I would decide to look at country songs produced over the past decade. But even this would be too much content. To further narrow things down, I could look at the top five selling country songs for each year for the past decade. Note, different kinds of media usually have a mechanism for determining what's most popular, through either sales, downloads, viewers, or circulation, so determining what's "most popular" is usually fairly easy. Moreover, because it's the most popular this means more people have seen or heard it, thus suggesting it has the greatest influence.

For the country music example provided here, I now have 50 country songs (five each year from the last decade) whose lyrics I can thematically examine. This is a manageable amount of material. I could then get the lyrics for each of these songs (at this point most song lyrics can be found online) and systematically read through them, categorizing or coding them according to their content. Do the songs talk about love? Do they talk about violence? Do they talk about a broken heart? Do certain themes appear alongside one another so that one song talks about all three of these themes at the same time?

When conducting a content analysis it is important to be clear about the units of analysis and levels of analysis. In this case, I am coding the lyrics to individual songs, which means I am looking at specific words in the songs, but then assigning each song to thematic categories which have inductively emerged during the course of my analysis. Sometimes the content is **manifest**, which means that it is in plain sight on the surface and everyone can agree on what the lyric means. For instance, if the protagonist in a song sings, "I'm gonna get my gun and kill you because you wronged me so," we would code the overall song under the categories of *domestic violence* and *love gone bad*. This is pretty apparent from the lyrics themselves, and most independent observers would likely strongly agree on this, which suggests that this coding is reliable. However, matters become trickier when some interpretation is required. This is **latent content**, material whose meaning is not clearly on the surface, but must be interpreted taking the overall context into account. For instance, a song protagonist might sing, "I love you so," but while this on the surface may appear to be an ode to a beloved, it could be sung sarcastically if the remainder of the song is about "how you wronged me so." In cases of latent content, to avoid subjectivity and bias it is up to the researcher to provide clear and careful guidelines on how to code the material to ensure both the reliability and validity of the results.

Like most qualitative approaches, content analysis is an iterative process of analysis, meaning there is a constant back and forth between material being coded and the coding categories themselves. One advantage of content analysis is that it is a forgiving technique: If you make a mistake coding or miss a category the first time around, you can simply go back and do it again without penalty. This is not the same as survey research, where if you make a mistake in not properly asking a question it could have adverse consequences for the results.

MIXED METHODS

These unobtrusive methods of collecting information are not exclusive to qualitative methodologies and can be used for quantitative studies as well. What makes the difference between the two forms of data collection is the design and analytic approach taken with the collected information. If we are interested in counting the frequency of some image or communication, then the study will be quantitative in nature. If, on the other hand, we are engaging in a qualitative analysis, we are likely to examine the material for common themes and motifs.

RESEARCH IN ACTION 11.1

ILLUSTRATION OF A MIXED-METHOD EXAMPLE

Source: Covert, Juanita (2003), "Working Women in Mainstream Magazines: A Content Analysis." *Media Report to Women, 31*(4) 5–14.

Hypothesis 1: Working women portrayed positively will be more likely to be described in terms of stereotypically feminine attributes than stereotypically masculine attributes.

Hypothesis 2: Working women portrayed as successful will be more likely to be described in terms of stereotypically feminine attributes than stereotypically masculine attributes.

Note that in using both quantitative and qualitative designs, the author employed both deductive and inductive types of reasoning. She deduced a couple of hypotheses from the existing theoretical literature on gender representation in the media. She also sought to induce patterns through a qualitative analysis of themes and topics that she found in the content.

The sample of magazines consisted of five popular women's magazine. They were chosen from the Folio 30 list that appeared in *Folio:* 'the magazine for magazine management' (*Folio*, 2001). I listed current magazines according to circulation figures. The five general-interest women's magazines with the highest circulation were included in the sample. . . . Two issues of each magazine were chosen randomly from the months of January to December 2002 for a total of 10 magazines. . . . The coding of these issues resulted in data for 449 articles (54 of which were for or about working women) and 1080 individual women (699 of which were working women).

Given that these magazines were the most popular, the author assumed (not unreasonably) that they had the most influence and were thus most representative of the phenomena. Note also how her sampling has narrowed a potentially overwhelming universe of content down to a manageable amount.

The author was the only coder for this study and a detailed coding protocol was created as a reference. A second coder examined one of the sampled magazines chosen at random to test the reliability of the study's measures. . . .

It is customary to check inter-coder reliability by having another person independently examine the same content and ideally come up with the same results. The higher the agreement, the less likely there is subjective bias creeping into analysis.

Within each article, the coder examined portrayals at the individual-women level. . . . If a women was coded as a working woman, other characteristics were coded as well. These included *good relationship/social skills**, *ambition*, *creativity**, *risk taking*, *intelligence*, *beauty/fashion sense* and being competitive* or *passionate**. Some of these relate to stereotypically feminine traits (*) and some to stereotypically masculine ones (Durkin, 1985; Basow, 1992). Two neutral attributes were also coded: being *hard-working* and having *good organizational skills*.

Note also how the author is clear about exactly what she is looking at (levels of analysis) and looking for (stereotypical characteristics) in terms of content. This clear operationalization allows another researcher to replicate the study and is a key feature of any quantitative design.

(Continued)

(Continued)

Results supported the first hypothesis. Working women featured in articles were more likely to be described using stereotypically feminine qualities than masculine ones (see Table 1). Twenty-two percent of women were described in terms of beauty or fashion sense. This was followed by 19% of working women described as good at building or nurturing relationships. . . . Results supported the second hypothesis. Successful women featured in articles were [also] more likely to be described using stereotypically feminine qualities than masculine ones. . . . [See Table 2]. Seventeen percent of successful women were described in terms of beauty or fashion sense. This was followed by 15% of [successful] working women described as good at building or nurturing relationships. . . .

[In terms of the qualitative themes which emerged from the research], Table 3 shows the percentages of articles that featured each of the topics coded. The topics featured most frequently were fashion and beauty (28% of articles), two traditionally feminine values. This was followed by 24% of the articles regarding struggle, and 17% regarding contributions a working woman made at (or through) her job.

As can be seen, the results are presented in numerical form and directly address the predictions made in the two hypotheses. Interestingly, the qualitative themes that emerged from the coding were also quantified and presented in a frequency table. Space precludes a full reporting of these results, which were also presented in a narrative form, but there was an extensive in-depth review and discussion of additional qualitative themes which directly addressed the two research questions posed at the beginning. Content from the magazines illustrating some of the key themes were reported in some detail.

Content analysis is also a good example of a type of obtrusive research that can employ mixed methods. As was discussed in Chapter 1, **mixed methods** refers to research studies that combine elements of both qualitative and quantitative methodologies. It has already been noted that different approaches have their own distinctive strengths as well as limitations. Many times, researchers specialize in either quantitative or qualitative methods. This specialization is often the result of a researcher's particular skill set and interests, and the type of research questions he or she decides to employ. Specialization is also a function of time because it is difficult to do everything at once! Nonetheless, while one may be more adept at using one methodological approach over another, quantitative and qualitative designs should not be seen in opposition to one another, but rather as complementary. By combining approaches one can gain the best of both analytic worlds, answering research questions in ways that are both broad and generalizable as well as in-depth and detailed. If time allows, you should experiment with using both quantitative and qualitative designs using the same research question, as a well-rounded approach will yield the most comprehensive results.

As an example of the benefits of a mixed-methods approach, consider the Research in Action 11.1 about communications researcher Juanita Covert, who was interested in how working women were being portrayed in women's magazines, and whether gender stereotyping was a factor in these portrayals. To address this topic, she performed a content analysis of women's magazines, using both quantitative and qualitative research designs.

Summary

The general aim of qualitative research is to provide a thick description of the behavior of members of a group and what it means within a sociocultural context. Qualitative data are often generated in the process of fieldwork, during which direct observations are carried out through a technique called participant observation. The role one takes during the course of participant observation can range from (1) a complete participant, to (2) a complete observer, to (3) some combination of the two. A complete participant is the researcher who is not simply observing a population or event of interest, but is also experiencing firsthand everything that participants are experiencing. A complete observer, on the other hand, is the researcher who is simply observing without participating or experiencing the dynamics that participants are going through. The participant-observer role combines elements of both the insider and outsider perspectives and balances the desire for conveying the subjective experience of group members with the scientific aim of understanding behavior within a broader analytic framework. While doing fieldwork the researcher will develop rapport with informants who will provide valuable information and contacts, as well as help the researcher generally understand what is going on. During fieldwork, the researcher keeps a field journal which contains detailed descriptions of the setting, as well as a forum to explore ethical challenges, consider methodological issues, and reflect on theoretical insights.

In-depth interviews are another way of eliciting information from research subjects. These in-depth interviews may be conducted either one-on-one or using focus groups. One-on-one interviews with participants may be categorized as: (1) structured interviews, (2) semi-structured interviews, and (3) unstructured interviews. Structured interviews have a clear interview protocol where all the questions are written and the researcher asks the same exact question to every participant. Semi-structured interviews are more flexible in terms of the questions asked and the follow-up with participants. While a specific structure and protocol exist, the researcher is clear on what the questions will be for every participant and is more likely to follow up in a question of interest and probe for more answers. Unstructured in-depth interviews are interviews that feel more like a conversation between the researcher and the participant.

Focus groups are a specific type of group interview where a few people are brought together to discuss a topic. Participants may fill in answers from each other, may be more confident to talk about something if another participant is also talking about it, or even have opposing opinions about something and are likely to share their point of view. Focus groups are cost-and-time effective, but are more challenging to manage and may require more than one researcher in the room to facilitate the discussion.

The product of qualitative research is usually an ethnography. Ethnographies are the written results from the perspective of an insider that provide a thick description of the group under study. They are widely used by anthropologists, but many other disciplines like sociology use ethnographies in their studies. Often, researchers combine more than one type of data collection to best fit the study. Typically, the researcher prepares an ethnography by combining a few methods of data collection, such as participant observation and in-depth interviewing. The qualitative research needed to produce an ethnography is labor intensive and usually requires more time than other types of design, but the richness of the outcome is worth the effort.

In addition to the fieldwork and interviewing, qualitative data can also be gathered by using unobtrusive methods. Indirect observations may be made by examining (1) the traces of human behavior or thought people have left behind; (2) archival data, which are formal records produced by institutions; or (3) secondary data that are collected by another researcher prior to your own study.

The various sources of trace data include (1) physical objects, (2) still images, (3) sounds, (4) moving images: videos, and (5) texts. Trace data in the form of text and images are very suited to content analysis. A content analysis is a very popular unobtrusive method of research which can be undertaken by the novice researcher. A content analysis systematically examines the trace materials described above. If one can effectively design a content analysis, one has effectively covered all the main points in the research process.

Using archival or secondary data is attractive because the data are already collected and a study will have lower costs and be conducted in timely manner. However, these two types of data collection can face problems with having a different sample or different type of information from the one that best fits our study. They are also prone to sampling and measurement errors.

Finally, we looked at mixed methods, which employ elements of both quantitative and qualitative research designs. One benefit of using mixed methods is that undertaking a variety of approaches to the research questions provides a fuller, better-rounded view of the phenomenon under study. Research employing mixed methods tends to produce results that are both broad and generalizable as well as in depth and detailed. To carry out a mixed-methods project one needs to develop a range of research skills rather than specialize.

Key Terms

Complete observer: the researcher who is simply observing without participating.

Complete participant: the researcher who is experiencing everything that participants are experiencing firsthand.

Content analysis: the systematic examination of the trace communications people have left behind.

Ethnography: a piece of cultural writing produced from the results of fieldwork that provides a thick description of the beliefs, practices, and overall ethos of a group.

Extended case method: a more deductive approach to theory development in fieldwork that compares the theoretical insights gathered from other ethnographic case studies with one's own fieldwork experiences.

Field journal: a written and bound collection of notes where the researcher records observations and insights during fieldwork.

Fieldwork: data collected while the researcher spends an extended period of time being a participant-observer with a group in its own natural setting.

Gatekeeper: the person (or persons) in charge of the setting where fieldwork is being conducted.

Grounded theory method: an inductive approach to developing theory by systematically comparing and contrasting qualitative observations to look for emerging generalizable patterns.

Hawthorne effect: a term, named after a famous study, used to describe how participants in a research study may alter their behavior simply because they are part of the study.

In-depth interviewing: a data collection technique in which the qualitative researcher elicits information from a respondent through a series of open-ended questions.

Informants: group members who provide useful information to the researcher during the course of fieldwork.

Interview schedule: the written series of open-ended questions used to conduct an in-depth interview.

Latent content: content whose meaning is below the surface and requires some interpretation in context.

Manifest content: content whose meaning is plain and clear on the surface.

Mixed methods: a study that combines elements of quantitative and qualitative research designs.

Participant-observation: a mode of direct observation employed during fieldwork.

Probe: an interview technique where the researcher asks a follow-up question to elicit more information from the respondent.

Rapport: the sense of trust and connection that develops between the researcher and the research subject during the course of fieldwork or in-depth interviewing.

Saturation: the point toward the end of fieldwork when the researcher is no longer taking in any more new information.

Thick description: a rich and highly descriptive account of human behavior that puts the meaning behind the behavior into a broader cultural context.

Traces: the cultural artifacts such as (1) physical objects, (2) still images, (3) sounds, (4) moving images: videos, and (5) texts left behind by people that can be gathered and systematically examined in an unobtrusive manner.

Unobtrusive data collection: a data collection technique where the researcher has no direct contact with any research subjects during the course of investigation, but instead looks at the traces that have been left behind by that population.

Taking a Step Further

1. How can we prepare for the perfect focus group?
2. What are some elicitation techniques we can use in qualitative studies?
3. How does a complete observer differ from a complete participant?
4. What are some ethical challenges we face in qualitative research, especially if we are a complete participant in a study?
5. What are some of the steps to prepare for a face-to-face interview on a sensitive topic?
6. How are case studies different from the other types of qualitative studies?

ABOUT THE CHAPTER CONTRIBUTOR

Charles Sarno is associate professor of sociology at Holy Names University in Oakland, California, where he has taught since 2000. In addition to qualitative research methods, his academic interests include the sociology of religion and deviant behavior. His current research and writing is on the Family Radio network, whose founder Harold Camping predicted the end of the world several times, most notably for May 21, 2011. His recent article "Church, Sect, or Cult? The Curious Case of Harold Camping's Family Radio" was selected for the 2016 Thomas Robbins Award for Excellence in the Study of New Religious Movements by the journal *Nova Religio*.

12

ENTERING, CODING, AND ANALYZING QUALITATIVE DATA

CHAPTER OUTLINE

WHAT WILL YOU LEARN TO DO?

1. Describe how to enter, clean, and organize qualitative data
2. Examine and code raw data
3. Analyze qualitative data

ENTER, CLEAN, AND ORGANIZE QUALITATIVE DATA

Like the quantitative information discussed in the last two chapters, qualitative information also needs to be sorted, organized, and coded before analysis can begin. Conceptually, qualitative information is handled in the same way as quantitative information, but some adjustments must be made because the data are delivered in word form rather than numbers. By its very nature, qualitative information will never be represented by numbers, but can still be systematic and well organized.

As you may recall from our earlier discussion, there are many types of qualitative studies, such as focus groups, participant observations, in-depth interviews, and case studies. Once all textual information has been collected, we need to closely read the data and follow rigorous analysis to find underlying meaning or possible relationships between events. Researchers have developed some effective and systematic ways to handle and interpret qualitative data, which we explore in this chapter.

In addition to the raw information we accumulate when conducting a study, it is essential to create a separate file of various records, including how many interviews/case studies/observations were conducted, when and where the accumulation of data points took place, and other information about the participants in the study. Each experience, whether an interview, focus group, or investigation of case studies, needs to be well documented. Basic information like the date, time, location, duration of the event, and participant characteristics should be logged.

The qualitative process requires attention to every single detail or dynamic surrounding an event, and at times we as researchers may not be completely sure of what is important. The key is to document everything that makes an impression during data collection or directly following it. In recording data, some researchers prefer to create name codes for the people they interview, while others assign numbers to participants. It may sound more realistic in a narrative if you use actual names, but remember these must be fictional due to rules of confidentiality.

Transcription

Transcribing an interview means writing down everything you hear from the recorded interview, without exception. Every utterance or nonverbal sound from the participant or the interviewer needs to be recorded. Naturally, this will take some time, but there is a silver lining. Because the process is so detailed, the same sentence may be heard many times before it is completely typed. This can slow down the process enough to allow for meditation and fresh thinking about the data. It also allows the researcher to notice characteristics of the data that may have escaped his or her attention earlier or to view the information from a new perspective.

Having a set of ground rules about transcription is helpful. A researcher needs to include everything that is heard, from grammatical errors and mispronunciations to background noise. We may be tempted to correct a grammatically incorrect sentence during transcription, but in the process, we may transform the interview and lose meaning, so it is crucial to write down everything just as we hear it.

Transcribing is a detail-oriented task that requires a lot of attention. It is a powerful process that allows for consolidation of thinking.

Some researchers hire someone to transcribe for them. It is a costly alternative, but it may be worth the saved time. If you do choose to do this, establishing ground rules with the transcriptionist will be especially important. With limited control over someone else doing this work, it may be hard to maintain your standards. Furthermore, if participants have revealed any personally identifying information during an interview, you cannot allow someone else to transcribe that interview without removing this confidential information.

RESEARCH WORKSHOP 12.1

GOOD TRANSCRIPTION PRACTICES

There are some rules on how to transcribe interviews in terms of minor details, such as how to indicate a pause or something that is inaudible. If you are conducting the transcription and analysis, you can create your own rules about how to indicate specific important details during your interviews. If you have hired a transcriptionist, it is even more important to provide a set of rules for him or her to follow regarding how the data need to be entered.

(Continued)

(Continued)

The following are some common rules:

(()) double parentheses may be used to indicate laughing, coughing, clapping, stomping, sneezing, or any other nonverbal emotions during the interview.

(.) may be used to indicate a short silence. You may have two different types to indicate uncomfortable silences and brief silences.

: may be used to indicate that a syllable is elongated.

{} may be used to indicate speech that is unclear and cannot be transcribed.

You may add notations according to other important verbal or nonverbal cues and adjust them to your own research.

Memoing

Reading the transcribed interviews is the first step of qualitative analysis. In fact, it is recommended that researchers read the interviews a few times and think about them at length in order to become actively involved with the written material (Creswell, 2013). This active reading occurs before the researcher begins the analysis and is accompanied by extensive memos and notes. Taking notes on what you are thinking as you read and reread the transcribed interviews is called **memoing**, which facilitates later analysis by bringing you closer to the written material. After all, the interview you are reading represents a moment in time—an experience that can become a living entity—especially if enhanced with comments, thoughts, shared experiences, and remembrances in the form of memos or notes.

Strauss and Corbin (2015) delineate the types of memos researchers create, but emphasize that the first goal is to allow unbounded thinking without worrying about categorizing the memos. Getting into the habit of writing about anything and everything, without consideration of how or whether it will be used, is the key to writing useful memos.

For example, memoing the Research in Action 12.1 interview was a crucial step in understanding how to make sense of the separate pieces of information that originally seemed unrelated. The process was quite straightforward. As the interview was being played repeatedly, the participant mentioned something about the turkey she bought the year before and how the turkey was 20 pounds. A quick memo that went something like "Interesting. She remembers how big the turkey was last year. Too much attention to detail," was quite useful during the analysis stage, as many participants showed great attention to different details regarding food.

Strauss and Corbin (2015) stress that classification of memos is a process that happens after the memos are written in free-thinking style. After the memos are grouped together, the researcher may attempt to classify them according to the information they provide. The following are some suggestions regarding how to classify memos, but any researcher can add categorizations depending on his or her study and the type of analysis selected.

A. **Theoretical memos:** These memos relate to a theory or make connections between the narrative and theoretical concepts. For example, using our Research in Action example, the researcher may create a theoretical memo of feminism and link all the passages from the interview that seem to hint at food restrictions.

B. **Self-reflection memos:** These memos connect the narrative with past experiences or simply note how the researcher perceives the written information. An illustration of these self-reflection memos could be linked to a portion of the interview where the respondent discusses how she goes online to search for forums of vegetarians because she is not well adjusted to her new diet. One personal reflection memo may be something like "Wow. I would have never done that. If I was not adjusting well to soy, I would simply switch back to meat. But I guess she took the commitment to become vegetarian very seriously."

C. **Hypothetical memos:** These memos propose or predict a relationship between different parts of the narrative, though it is still too early to come to a conclusion. An example of hypothetical memos would be in one of the interviews where the participant mentions that she loves Krispy Kreme donuts, but does not buy them because she is aware of their impact on her health. She also mentions how she became wiser about food as she grew older. At this moment, the researcher could write a hypothetical memo stating, "I wonder if this is going to be the overall theme for the rest of the interview. Is she evaluating her food behaviors over time?" This may or may not be the case, but since this is relatively early in the interviewing process, there may be a benefit to creating hypothetical memos.

RESEARCH IN ACTION 12.1

ILLUSTRATION OF AN INTERVIEW TRANSCRIPT

This is from an actual interview in a mock study conducted by the author of this book focusing on the relationship between food and gender.

I: *Alrighty, eee I am doing this study on eee ed. . . . how educated women see food and the meaning of food for them, eee . . . and how that has changed over time. Eeee . . . I wanted to thank you for volunteering in my study. It's a very small study on this . . . eee. . . . food behavior I want to say, but I don't want to mention anything, I just want to let you see how, how you see food, and what food means to you, and how you perceive it in . . . in your, in your everyday life. Eee . . . Do you enjoy food?*

A: *eee . . . yes . . . I enjoy food very much. I love eating*

I: *ehe*

A: *. . . I . . . so yes, I enjoy food very much.*

I: *eee . . . can you tell me a little bit more about that?*

A: *eee . . . sure . . . eee . . . let's see. . . . I . . . I like eating a lot but . . . not a lot of food I mean I enjoy eating*

I: *ehe*

A: *but I also try to balance that with healthy behaviors.*

I: *such as . . . ee . . .*

A: *such as . . . eee . . . well . . . I try to do like portions . . . I try to focus on portions. . . . so I eat pizza and I eat Chinese food, and I eat all different types of food just because I love eating, but I try to balance that with portions, so then, instead of having four five slices of pizza*

I: *ehe*

A: *I will have two. And sometimes I will take off the cheese.*

I: *I see*

A: *So, you know . . . eee it's ee . . . so it's like a compromise for me. I am able to enjoy it*

I: *ehe*

A: *different types of food and at the same time . . . I am . . . I put a limit on myself. . . .*

I: *ehe. How, how important is food in your life, like . . .*

A: *hm . . . let's see . . . {long pause}*

I: *you know some people feel like they can't live without food, or they think about food all the time, or . . . do you cook? I mean . . .*

A: *hmm . . . well . . . let's see . . . well . . . I don't know if I think about food all the time. I wouldn't say that I think about food all the time, but . . . I make it a point to eat often during the day, because I feel for me it makes a big difference.*

I: *ehe*

A: *If I don't eat often . . . so I notice the change in my behavior and . . . in my mentality. . . . So . . . you know . . . I feel like I can't focus as well and I feel like I am dragging if I don't eat often. So then . . . I just make it a point to have . . . during the weekday . . . eee . . . five or six meals . . . small meals . . . a day . . . and that . . . that keeps me . . . feeling good.*

I: *OK. Eee . . . you didn't tell me whether you cook or not . . . ee . . .*

A: *eee . . . yes . . . I . . . let's see . . . well . . . before the program started I cooked . . . before the PhD program started I cooked more often. . . . And . . . now. . . . I . . . don't cook . . . I would say . . . I don't cook most of the week.*

I: *ehe.*

A: *So . . . you know . . . we . . . on Mondays and Wednesdays . . . which are our late nights . . . I go to Taco Bell . . . in the evening . . . like after class . . . and I eat my . . . eee . . . you know . . . like one burrito . . . eee . . . on my way home . . . and the rest of the time . . . we eat out . . . and on Tuesdays and Thursdays . . . except Tuesdays and Thursdays which I . . . we usually eat at home . . . just something that I prepare . . . so I cook . . . maybe two out of seven days of the week . . . and I . . . I also cook baked goods . . . like . . . during . . . during the weekend. I cook . . . like muffins for example . . . and I use that . . . I bring them . . . to . . . here, on campus. And those are my snacks.*

I: *I see . . . eee . . . is that something that you like more about cooking? What is . . . what is that you like more about cooking?*

A: *eee. What I . . . one of the main things that I like about cooking is that I can . . . I can control ee . . . ee . . . healthiness of the meal, so I know what I will be eating is healthy, just because I picked it and I know I know how I cooked it . . . so I like that part. You know, I make sure to have a balance of different things, like vegetables, beans, ee . . . you know, brown rice, but if you go out to the restaurant then they might not have a brown rice option and you have to do white, so you compromise more when you eat out than like, when you cook for yourself, so that is one of the things that I like about cooking.*

I: *Can you tell me some of your favorite meals?*

A: *eee . . . favorite meals . . . let's see . . . that's a good question . . . eee . . . I can say favorite . . . favorite food items . . .*

I: *OK . . . well*

A: *can we start with that maybe?*

I: *Yeah, sure.*

A: *Lately I like chickpeas a lot. Cause we just started eating a vegetarian diet several months ago*

I: *Oh, I see.*

A: *so . . . chickpeas I like a lot. Eee . . . you may know being a vegetarian . . . that beans are a staple for a vegetarian diet, so that's one that I like a lot . . . and . . . ee . . .*

I: *What do you do with chickpeas?*

(Continued)

(Continued)

A: *What do I . . . let's see . . . sometimes I cook them on the stovetop and other times I just . . . I eat them right out of the can like mixed in the salad. So, chickpeas is one, and another one is . . . now with the vegetarian diet I am eating a lot more fruits and vegetables. And It's . . . I am really happy about the change in in the diet . . . you know the change to vegetarian eating because we are eating more healthy now, so that's really why I like it, and one of the greatest things is that I notice a change in the behavior and also ee . . . functioning . . . and ment . . . mentality . . . so*

Transcribing this interview was quite challenging because each interview lasted approximately 30 minutes, with 10 such interviews from different women of different ages. These women came from walks of life and each one had a unique view on food, meals, and preparations. However, it was during the transcription process that the researcher noticed various differences between participants' habits as well as a few common themes: (1) food was conceptualized as relating directly to health, (2) food was so important that even small details about food were memorable, and (3) food was a part of their personality. Whether it was a young girl who did not cook at all, a vegetarian, or someone who was dieting, all participants had a close, attentive relationship with food. They remembered details about food, were aware of the impact of food on their health, and followed specific rules, albeit different from one another.

Note how every single pause, every little letter uttered by the interviewer or the interviewee is recorded in the transcript. Sometimes the interviewee is nodding (recorded as "ehe") and at other times she or he is simply thinking or pausing ("eee"). Depending on the context, this could mean hesitation to answer, time to think, or could even indicate a manner of speech. It is the researcher who decides the importance of these seemingly unimportant utterances.

About computer-assisted qualitative data analysis software: There are various options to code and analyze qualitative data, such as Nvivo, Atlas.ti, RQDA package of R, text-mining of R, HyperResearch, and Dedoose. They are useful tools to help the researcher organize, clean, and look at the data from different perspectives. These software programs allow for clustering codes, creating customized diagrams, identifying themes, and interpreting data more effectively. Most software programs have trial periods varying from 15 to 30 days, and prices can vary with the exception of the RQDA and text-mining packages of R, which are free environments. Most universities can sell these programs at a discount, so it is helpful to ask about their availability from your library or information technology (IT) department.

Diagrams

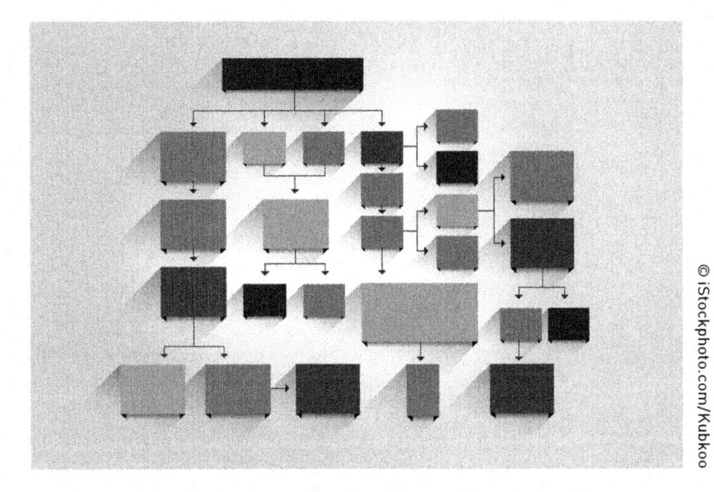

Creating diagrams is a helpful process that will crystallize relationships between constructs.

We live in a visual world where the expression "a picture is worth a thousand words" is truer than ever. Since many of us are visual learners, expressing our thoughts in graphs and diagrams is sometimes more helpful than other types of analysis. We tend to remember information better when we visualize it. Diagrams are not only useful for record keeping, but can also lend a systematic and organized view of the data (Corbin & Strauss, 2015). The most basic way of creating diagrams is by using a pencil and paper while we read the qualitative information and are struck by a new idea. Various software programs allow researchers to create a range of diagrams, from simple to complex. We may use diagrams in various phases of our analysis to put things into perspective, to know where we stand in the analysis, and to determine what may lie ahead.

CODING QUALITATIVE DATA

When presented with written information from an interview, a focus group, or an observation, we find ourselves searching for the underlying meaning. To find meaning, however, we need to organize and categorize the narratives we have collected. **Coding** is one way of bringing order to this written information. A code is a word or a few words that capture common meaning or categorization. The process of coding involves aggregating the text or visual data into small categories of information, seeking evidence for the code from different databases being used in a study, and then assigning a label to the code (Creswell, 2013, p. 184).

In other words, we may read something like the following paragraph (codes created for this paragraph are in parentheses and the words they correspond to are underlined): "*yes that's true, we've had a couple of barbecues and . . . they do make a point to either get salmon or some other fish for us (types of food), or a veggie burger, a black bean burger (types of food). yeah. . . . It does feel a little weird though going to barbecues . . . and . . . . knowing that the same grill where your fish is cooking there is a hamburger coming, you know, from*

RESEARCH IN ACTION 12.2

DIAGRAMS

Source: Wagman, P., Björklund, A., Hakansson, C., Jacobsson, C., and Falkmer, T. Perceptions of work-life balance among a working population in Sweden. *Qualitative Health Research, 21*(3): 410–418. First published date: August-03-2010 10.1177/1049732310379240.

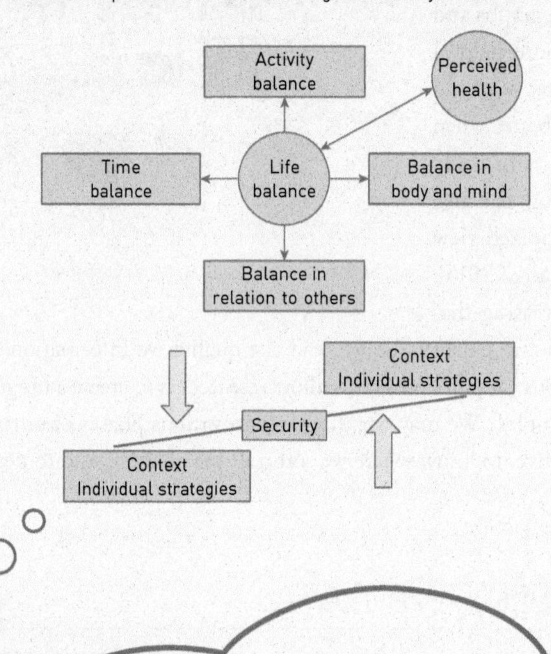

The researchers conducted face-to-face and phone interviews with 19 individuals (12 women and 7 men), with a range of ages from 26 to 64 (mean = 48). This was a heterogeneous group in terms of marital status, education, parental status, and employment fields. After transcribing the interviews, reading them multiple times, memoing, and coding, the researchers distinguished four dimensions of life balance. You can see these four dimensions in the diagram shown around the circle (life balance). Activity balance was defined as work-life and work-family balance, but also included mandatory or free activities. Balance in body and mind related to physical activities, sleep, and eating habits. Balance in relation to others included the notion of oneself and the social environment. Time balance was about the actual time awareness.

Also, as you can see from the second half of the diagram that there is a sense of security that came with having balance between work and life, and a disruption of security when there was no balance. Clearly, the diagram depicts these issues in a neat and immediately understandable way compared to explanations in words.

a cow (uneasy feeling/reflection for meat eaters). And that is kind of weird to think about that, but then I realize at the same time that I used to eat this way like that less than a year ago (thinking back in time about food). So, I mean, I don't want to sound hypocritical (self-judging). I just think about it sometimes, not all the time, the fact that they are in the same grill (uneasy feeling/reflection for meat eaters)."

As you can see, the same code "types of food" was attached when the respondent mentions fish and when she talks about the veggie burgers. Also "uneasy feeling/reflection for meat eaters" appears twice in this paragraph when the respondent mentions how the fish she eats is being cooked next to a piece of meat. Simply put, coding is a personal interpretation of the words in front of us. Some researchers may read the same sentence and not quite perceive the conversation as "reflection for meat eaters" and others may see even a form of judgment of meat eaters. Because we all have the tendency to interpret things differently, qualitative studies often use a team of two or more researchers to go over the same text and code separately. This way, we can strengthen the validity (review Chapter 5 for this term) of the study.

As you recall from Chapter 5, semi-structured interviews allow participants to elaborate on their thoughts and even allow for the interviewee to follow up with new questions. One big advantage of the semi-structured interviews is the ability to ask every respondent the same set of questions. This feature is quite beneficial when coding because we can quickly determine where similarities are found. When our interviews are unstructured, we start coding the information by grouping together similar and different things.

In the previous coding example, we can code as "type of food" every single time a food is mentioned by this participant and others. This can be very helpful when we compile a long list of food types mentioned during the interviews and compare them by gender (if that is the focus of the study). Coding is an attempt to classify and categorize so we can interpret and find meaning in the data. Coding promotes further analysis because the researcher is thinking about the meaning of the words during this process.

Two basic types of coding are available as we begin organizing our written material. First, there is coding as we see it in the writing, or **in vivo coding**. An example of in vivo coding is if we see a word or expression in the written text that seems important, like "vegetarian." Until we read the interview, we were not thinking of vegetarians per se, but this may be important, so we create the code "vegetarian" and apply it every time we read something related to it in the interview. Second, there are **preexisting codes** from the literature. These preexisting codes are often created before we start coding by making a list of things we are hoping to find in the text. For example, our initial thinking before reading the interview was that women are very mindful when it comes to the types of food they eat, so we created a preexisting code called "attention to food." Now, we read the text and if we see something that seems to refer to the participants' attention to food, we use the preexisting code.

Most researchers use both in vivo and preexisting codes simultaneously. Though they have already created a list of codes they are looking for, they also obtain new ideas from the material they read and create codes as they go. After the researcher has completed the first round of codes, he or she will begin reading all the codes and the written information to be analyzed in an attempt to make sense of it and to distinguish patterns.

At this stage, the researcher groups the codes into larger categories. Depending on the study, there are different ways these groupings occur, but some common grouping codes are (1) similarities/differences in describing the same phenomena—for example, attention to food; (2) similarities/differences in the stated frequency of a concept or topic, such as the types of food the participants are mentioning; (3) similarities/differences in the expressed attitude regarding a similar problem or situation—examples could be the attitudes toward meat eaters if the participants are vegetarians, or if they have other types of biases; and (4) similarities/differences in the cause-and-effect patterns of a problem or phenomena, such as the relationship between food and health.

Finally, these grouped categories can be further connected to one another in even broader groups called **themes**. Therefore, from an initial, narrow perspective of generating individual codes, we may then generalize more broadly to data categories and then even further to data themes. In this way, the interpretation of results will be much simpler and conclusions more easily drawn. This process builds on the earlier efforts of the researcher as he or she analyzed, pondered, or critically examined the data. In the illustrations used throughout this chapter, the attention to details when it came to food was a recurring theme for all participants. One participant remembered the weight of the Thanksgiving turkey last year, while another discussed in detail when and how a specific type of fish was on sale at the supermarket, and yet another described how she converted her regular diet to one that is gluten free.

ETHICAL CONSIDERATION 12.1

CONFIDENTIALITY

The best way to protect the confidentiality of your participants occurs at the moment of interview transcription. Sometimes participants will reveal personal information that can make them vulnerable to identification. It is not sufficient to remove their name and basic information. For example, the participant in the interview example provided information on her origin (Cuban), her husband's name (Don), and the fact that she was in a PhD program at the time of the interview. None of this information is crucial to the study's goal, but all of it together can reveal her identity. While the interview was being transcribed, all identifying information was modified.

ANALYZING QUALITATIVE DATA

The process of analyzing qualitative data happens primarily inside the researcher's mind. It is exactly this quality that renders a qualitative analysis not only rich in depth of process, but subject to possible criticism. One's thought process will vary depending on personal experience, background, and point of view. While researchers have attempted to come up with many systematic ways to handle qualitative data, the analytical process can still be fraught with difficulty because each researcher is as unique as his or her study. However, qualitative analysis, when conducted correctly, should correct for those individual differences and bring out the best possible interpretation, akin to a unique work of art.

Research is, in fact, a work of art, even though it follows strict rules and regulations. Like works of art, research follows strict rules and regulations, but can still preserve the vision of the artist. This is true for quantitative analysis, but even more so for qualitative studies. Saldana (2013) has extensively written about the types of qualitative thinking, and for the purpose of this textbook, it may be helpful to become familiar with some ideas on how to "think qualitatively."

Recognizing Patterns

Patterns are all around us. As humans, we follow patterns from the moment we wake up in the morning to the moment we go to sleep at night. We have routines for everything, even the way we study is part of a personal pattern. Looking for similarities in actions, words, events, or feelings is one way of investigating qualitative work and what Saldana (2013) calls "thinking patternly." For example, sometimes all the participants interviewed refer to something by the same nickname or they make the same association about a particular event. These are the patterns that start to emerge in front of your eyes as you analyze the data.

Recently, a colleague of mine and I were interested in understanding the differences between an *applied sociology* program and a *pure sociology* program. There are only 16 programs in the United States, undergraduate and graduate together, that call themselves Applied Sociology. We contacted them and asked a few questions about their program—how it was different from a sociology program, whether there were changes in the courses students take, and other similar inquiries. Luckily, each program answered all of our questions. Once the written text was ready and available, we started coding independently and then compared the results.

We checked for differences and similarities to gain some understanding on the reasoning behind changing the program name. Our codes revealed a great deal of information that we had expected, such as the strong links to community service and the attempt to apply sociological skills in the workforce, but there were other findings that surprised us. Almost every program mentioned they had no resources to market their program and all the marketing that was needed was done by professors. This pattern emerged, case by case. We saw that almost every department had similar practices when it came to marketing and almost all mentioned the lack of resources. Thinking in patterns was quite beneficial to this ministudy.

Creating Categories

When it comes to qualitative work, the information may seem scattered, with various data pieces all over the place like a messy dorm room. The key to cleaning it up is to first discard everything that is not necessary or useful, and then organize your objects according to defined categories. Instead of being spread out around the floor and desk, current books need to be placed in a bookcase and old unnecessary books should be discarded. The same rule of thumb is applied to all other items in the room until we have a clean, organized space. Qualitative work is similar to organizing your room—it is a way of producing order from disorder.

Having so much written information in front of us is daunting, to say the least. At the start, coding involves removing extraneous information while keeping the information of value. Then, in the analysis phase, we assemble similarly coded information and organize it into larger categories so we can see how different participants approached the same topic of interest.

Let us use an example of a researcher who is interested in understanding how teachers recognize students with behavioral problems. This researcher may have conducted many different interviews, in addition to observing participants in classroom settings to understand how teachers distinguish between a typical child and a child who is exhibiting behavioral problems. During the analysis, we begin by grouping together all the answers that teachers provided when asked directly about their process. In this example, all the teachers were asked to describe the details that make them certain about a child's problematic behaviors. We may start our work by grouping these answers together without thinking of any patterns yet. We simply want to see how these teachers answered this very same question.

In this instance, we are organizing their answers by a specific category that we have created from the moment we created the question. We asked them to tell us how they know when a child is exhibiting behavioral problems and some illustrations to these scenarios. Some teachers describe children who cannot sit still and constantly run around. Others point out children who consistently forget their homework. Still others pay attention to whether the child wanders during lecture time in class. Even more answers can come from teachers who watch to see if the child fidgets or exhibits uncontrollable behaviors. Although these answers may seem like they do not relate to each other or show any pattern at all because the behaviors mentioned are unique to that teacher, we can organize them together under the same category: that of how teachers perceive children in their classes. After all, we may be interested in this discovery of how teachers may consider different children as being problematic.

Searching for Frequency

Although frequency seems to fit into quantitative analysis, it plays an important role in qualitative work as well. Often we may find that a specific word, or a group of words, is repeated many times by different study participants. As we discover this during coding, we may need to return to this feature in the analysis process because the fact that a topic or concept is repeated over and over may be meaningful to the entire work. It may even indicate something we were not aware of before. The beauty of qualitative analysis is truly in finding these powerful details that are later brought to light for the analyses.

To easily recognize expressions or words that are most frequently used by participants, we may employ various software packages. We can even visualize these words to illustrate our work. Figure 12.1 shows an example of a Wordcloud library using a dataset of Hillary Clinton's emails from kaggle.com. Kaggle is a website that includes many free datasets where people work individually and share their work on various analyses with others. In this example, Clinton's emails were downloaded as text to find the most frequently used words. Then, the software was used to map these words visually. Instead of using a histogram, these words are mapped based on their size for a more powerful illustration.

Considering Diversity

Saldana (2013) talks about "thinking multiculturally" to bring the attention to our diverse world and the many elements that we need to keep in mind when analyzing data. If we talk to parents about their childrearing practices and we encounter comments that may strike us as odd or strange, we may want to consider an underlying cultural explanation before we categorize an answer as "strange." Our

Diversity in thinking is a powerful tool that allows the ability to think differently about problems of the society.

FIGURE 12.1 ■ Output Script of Hillary Clinton's Emails

world is diverse not only in terms of ethnicity and country of origin, but also in terms of gender, sexual orientation, religiosity, income, education, as well as many other characteristics. As researchers, we need to keep an open mind about everything and consider all possibilities. Since diversity is part of the world we live in, our continued awareness of it will result in better studies, better investigation, and much better results and discussions. We need to understand the advantages of diversity not simply from a human rights standpoint, but most importantly from understanding the diversity of thinking. Various backgrounds and exposure to different circumstances give the researcher new ways to look at the world and, as a consequence, new solutions. Diversity today is much more than the right to include individuals from different backgrounds. It is the ability to have different perspectives, be able to think differently, and offer new ways of looking at the world. Understanding diversity in terms of thinking is a crucial step to produce a higher quality of research studies that will ultimately improve our society.

Remembering the Past

There are many cases when past experiences, personal history, or other types of background will influence participants and impact their responses. Thinking about your participants' history encompasses more than a traumatic event or a major life occurrence—it can also be very simple. For example, we may be interested in investigating adolescents' likelihood of involvement in risky behaviors, such as alcohol and drug use. As we read and code the written interviews, we may want to consider the parenting styles used in raising the adolescents or their parents' involvement with risky behaviors. Could it be that children of parents who are involved in risky behaviors simply perceive such behaviors differently from other children?

Before considering whether risky behaviors relate to school performance, popularity, or family income, we may also analyze the background element to see if that is one factor among others for such behaviors. Taking into account as many elements as we can when analyzing data will result in a higher quality of data analysis and a stronger study.

Thinking Theoretically

From the moment we begin our formal education, we learn about different theories. We remember some of them and forget others. In the case of research, you will find the most applicable theories and apply them to your study.

The process of thinking theoretically is the opposite of the desire to apply a newly learned theory. This process sparks appreciation of and connection to a theory when we read about a particular social interaction or a current event. We might become excited because a "feminist viewpoint," a "Freudian perspective," or a "functionalist standpoint" suddenly makes sense and helps us to interpret the information we have just collected in a meaningful way. There are countless theories available and, of course, we will not be able to know and apply them all. However, we will be able to think about some of them as we conduct our qualitative analysis and they will help us to make necessary connections between our material and a larger theoretical context. It may help to refresh your memory of some key theories and keep them in mind as you read transcribed material from your qualitative research.

Thinking Critically

The ability to think critically is a skill that is developed over time by exposure to different ideas. It is the ability to look at something from multiple perspectives and to challenge our usual ways of thinking. This concept is abstract and subject to interpretation. However, here is a bit of a truth—feeling confused, unclear, and a bit in the dark about how to proceed is an important part of the analysis process in any research study. If we are fully clear and certain of our direction from the beginning, then there will be no learning process; we are simply accumulating knowledge.

However, in the active process of analyzing our data, we need to apply critical thinking skills, which will not happen without a degree of initial uncertainty and confusion. It seems contradictory, but without mixed feelings and even negative thoughts, we are not likely to realize brilliant interpretations of our work. In fact, the process resembles a rain shower after a sky filled with darkness and clouds; our analysis will pour down beautifully if we first endure an uncertain, cloudy, and circuitous process of considering various interpretations and applying critical thinking skills to them. In this process it is best not to stop and question your thinking because self-doubt, though understandable, will interfere with your creativity and critical thinking potential.

For some researchers, thinking critically can lead to a point of uncertainty about everything. There are certain times in life when you simply should not overthink. You can think of day-to-day examples, such as when you're shopping for a pair of pants for yourself. You will want to consider how they fit as well as the price. But what if you start wondering about the following points:

- What if they don't fit after one month because I am trying to lose weight?
- Does this color match my shirts?
- Do I look better in the other pair of pants I tried on?
- Can I get these pants online for less?
- Is this material durable?

We all have had these moments and know the result well. We usually leave the object in the store and walk out empty-handed. When we overthink and reconsider so many possibilities, it leads to confusion and frustration. However, when it comes to research and interpreting our findings, we welcome this type of confusion because it encourages productive critical thinking.

Summary

Analyzing qualitative data is different from quantitative analysis. We attempt to use statistical knowledge to analyze data in quantitative studies, but rely on different principles when analyzing qualitative work. Qualitative studies often investigate a few participants, with a lot of lengthy and detailed information. So how do we make scientific sense out of these studies? A lot of the analysis in qualitative studies involves the deep thinking of the researcher. Before we try to analyze qualitative information, the raw information is transcribed and coded.

Transcribing an interview is to write down everything that is said (and unsaid) without exceptions, and involves listening to the same sentence repeatedly. Once the data are transcribed, the creative process of coding begins. Most researchers also take notes on the written narrative—a process called memoing, which is considered another crucial aspect of qualitative analysis. There are many types of memos depending on the researcher and the study, but some common types include theoretical memos, which involve linking the narrative to theoretical ideas; self-reflection memos, which connect the narrative to personal experiences; and hypothetical memos, which propose or predict a relationship between different parts of the narrative.

Researchers have identified various types of coding, such as in vivo coding and thematic coding. With in vivo coding, we code in the narrative from the exact words used in the transcribed material. Thematic codes are preexisting codes created by the researcher. Although there are various ways to categorize coding and memos, it is important to remember that researchers will either have a second person code the same information or find other ways to strengthen the study, because we are all subject to different interpretations. The process of memoing and coding involves deep thinking and interpretation. After the process is over, the researcher may attempt to classify the types of codes used.

Recognizing patterns is the first step in analyzing qualitative information. Researchers are searching for information that emerges repeatedly. They also look for related topics among participants. Finding patterns is one way of making sense out of the qualitative information. Diagrams are used to visualize the information and make sense of the researchers' thoughts. Sometimes, mapping out an initial idea can help the researcher to gain a different understanding of the problem. Other times, a diagram crystallizes parts of a researcher's thinking that were not quite clear in his or her head.

Thinking is a crucial part of the analysis in qualitative work. Some broadly used forms of thinking include considering diversity, remembering the past, thinking theoretically, and thinking critically. Our world is diverse in culture, ethnicity, race, sexual orientation, religion, political affiliation, and so many other ways. Being able to consider different perspectives while reading written material is a valued skill in qualitative analysis, which can add rich information. As important as diversity, the background or history of participants must be weighed. Past events often shape who we are, what we think, and how we act. Omitting our histories will result in an incomplete analysis of the data and a lack of depth in analysis.

Being able to apply the theories we've learned happens almost naturally, since humans tend to categorize the new knowledge they encounter. One way of making sense of new information is by putting it into a predefined category. Linking new information to previous theories is one way to make sense of our world, and an important feature of qualitative analysis. Finally, thinking critically is our ability to think outside of the box. It is not a skill we are born with as much as a skill that we learn and improve on throughout our lives. Critical thinking means considering and reconsidering everything we know and assume. Questioning everything we read is a great way to practice critical thinking.

Key Terms

Coding: bringing order to written information in which text or visual data are categorized into small portions of information.

Hypothetical memo: memo that proposes or predicts a relationship between different parts of the narrative.

In vivo coding: coding that is created as it is seen in writing.

Memoing: taking notes on thoughts while reading.

Preexisting codes: codes that are created before the coding begins.

Self-reflection memo: memo that connects the narrative with past experiences or that provides notations on the researcher's perceptions of the written information.

Themes: broad groups of categories.

Theoretical memo: memo that relates to a theory or that makes connections between the narrative and theoretical concepts.

Taking a Step Further

1. Can you illustrate *theoretical memos* with an example?
2. What is the role of thinking in qualitative research and what are some of the most widely used types of thinking?
3. How do we find patterns in qualitative studies?
4. What is the process of coding for qualitative research and how is that different from coding in quantitative research?
5. What is the role of diagrams in qualitative studies?
6. What is the difference between a self-reflection memo and a hypothetical memo? Illustrate with examples.

13

RESULTS AND DISCUSSION

CHAPTER OUTLINE

WHAT WILL YOU LEARN TO DO?

1. Write a results section
2. Describe how to present results in quantitative studies
3. Explain how to present results in qualitative studies
4. Prepare visual representations of results
5. Compose a discussion section
6. Develop a recommendations section based on methodology or topic

REPORTING RESULTS

The results section is perhaps the most straightforward section of a research paper. It represents a complete report of your findings and it follows a basic structure. Presenting your work in writing can be intimidating at first. Sometimes a blank page stares back at us, and we are at a loss not only on how to begin, but where to go from there. Anything we write feels clumsy and not quite representative of the work we put into the study. Rosnow and Rosnow (1992) proposed that the best way to write is by starting with the section that feels most comfortable. Then, the researcher can expand, think, analyze, converse, and edit as well as add to the other sections as the work progresses. Other authors like Cloutier (2015) also agree that writing anything at all is a good start, because it gives you a starting point and information that can be revised and transformed even if all of the text is reconstructed from that point on.

Demographics Section

The results section generally begins with the demographics of your sample. Demographics are the basic characteristics of the population, such as gender, race and ethnicity, age, height, weight, marital status, sexual orientation, religious affiliation, employment status, and income. You will describe the population by providing these key characteristics starting with the number of participants in the study.

The results section is all about being as descriptive, detailed, and exact as possible. Your goal is to familiarize readers with your study, keeping in mind that they have no idea what the study is about or who participated in it. No detail is irrelevant and the information should be presented in such a way that readers can easily put each part of the study in context. For example, when reporting the number of women and men in the study, choose percentages rather than raw numbers because they serve as better descriptors. If you present the raw numbers, you are asking the reader to think about the numbers and put them in context rather than presenting your findings in the best light possible.

Demographics give us a summary of the basic characteristics of the participants in our sample.

Consider these two examples:

Example 1

There were 79 participants in this study. Fifty-eight were women and 21 were men. Forty participants reported coming from single-parent families and 39 reported coming from two-parent homes. Specifically, 29 women and 11 men came from single-parent families, which was an added advantage for this study.

Example 2

A total of 79 participants volunteered for this study. A great majority, 73% ($N = 58$) were women and 27% ($N = 21$) were men. However, half of the sample, or 50.6% ($N = 40$), reported coming from single-parent families. A great advantage of this study is that 50% of women and 50% of men were raised in single-parent homes. This distribution allows for a comparison between men and women regardless of the low number of male participants.

In Example 1, the answers are reported simply and paint a more complex picture for the study. Even though there is a statement saying that the distribution is advantageous, it is difficult to understand why that is the case unless the reader takes out the calculator and starts counting the number of participants according to gender and the state of the families who raised them. We know that asking the reader to do additional work besides reading the information may be confusing. In the second example, the same information is reported for the reader, but it is clear why the distribution is beneficial for this study. The percentages put the information into an easier context to understand. Now the reader can comprehend and remember all the details.

The demographic information is almost always important for the study even if the focus of research is somewhere else, seemingly unrelated to demographics. Basic characteristics such as race, ethnicity, gender, and income are often the hidden causes of further findings in the study.

© iStockphoto.com/Rawpixel

Parenting practices are so crucial to development of children.

Sometimes, a study makes bombastic claims and receives a great deal of attention from the media, but we are somehow unconvinced of the results. I once encountered this from a local organization that was attempting to collect information regarding the individual parenting practices of the parents of preschoolers. It had collected almost 500 responses to its survey when it found out that the majority of Caucasian parents read to their preschool children every single day. This finding was not replicated for the small subsample of African American or Hispanic parents. I became curious and visited this center to find out more. I did not believe that race and ethnicity could be so closely related to how much a parent reads to a child. Something else was missing, but I could not figure out what.

The person in charge showed me the survey and findings to convince me that this was in fact the case. Looking at the data, I noticed that parents were not asked about their education level or income level. The entire study compared different people from different levels of income and education, so the findings had to be taken with a grain of salt. The literature shows that the more education a parent has, the more likely it is that this parent will read to a child. Previous studies have also proven that a higher income translates to more parental resources, such as access to books, knowledge, and time. Without looking at these basic demographics, we cannot arrive at the conclusion that African American or Hispanic parents read fewer books to their children. Now, comparing African American parents and Caucasian parents of the same level of education and income is a different situation. In cases where demographics are central to a study, it may also be helpful to illustrate the information with a figure or table.

Findings Section

After reporting demographics, the results section describes findings regarding the key variables. Researchers typically provide general information about the variables using measures such as frequency, averages, or percentages. This is particularly important to the reader because it provides a broader picture of what the researcher found to be of interest.

Defining frequencies is the first step to draw a picture of our findings.

For example, a researcher may be interested in the relationship between reading to babies and their ability to speak. The researcher collects data and information on the reading habits of 100 families. When reporting the results, the researcher would provide descriptive information about the reading habits of these families before reporting a relationship between the children's ability to speak and parents' reading time. Information regarding the number of families who generally read to their children, along with the length of their reading time, will provide a valuable backdrop for presenting the findings of the study.

Once the initial information on the participants is written, the results section continues with findings that often relate to the relationship between two or more variables. If we were conducting a quantitative study, we would need to revisit our hypothesis to determine if the findings reject or fail to reject the null hypothesis. If we were conducting a qualitative study, we would report the relationships we have discovered in the course of the narrative.

RESULTS IN QUANTITATIVE STUDIES

Quantitative research often involves testing one or more hypotheses. We strive to investigate a relationship between variables and often predict how the variables will behave. Our prediction, as you know from previous chapters, depends on the literature we have read and on our theoretical perspective. In fact, we have already established expectations about the relationship of one variable to another. Once we have run our analyses, we will have a good understanding of whether our variables are actually correlated.

We are also aware of the strength and the direction of the relationship between variables. We know whether the variables of interest are associated with one another and whether one variable decreases or increases as the other increases in units. A straightforward statement of findings is customary, and, while it might initially seem bland and repetitive, you should report your findings in a direct and matter-of-fact style. Tables and figures are strongly recommended because they allow a visual presentation of detailed data.

There are two other scenarios that we must discuss regarding the results section: (1) rejecting the null hypothesis and (2) failing to reject the null hypothesis.

Rejecting the Null Hypothesis

To refresh your memory, the testing in quantitative studies relates directly to the null hypothesis. The null hypothesis states that there is no relationship between variables and stands in opposition to our predictions about the results. We engage in a relationship with the null hypothesis, even though we have an alternative hypothesis, that is, something else that we are trying to predict. But our alternative hypothesis is simply a prediction, so our first job is to attempt to reject the null hypothesis. If we manage to reject the null hypothesis and find a relationship between two variables of interest, we can further assume that our alternative hypothesis is a possibility, which we discuss later in this chapter.

Being able to reject the null hypothesis, however, may not mean that we have proven the alternative hypothesis. In fact, there are often scenarios in which we reject both the null hypothesis and the alternative hypothesis, because the relationship between variables is different than predicted. For example, a study exploring the relationship between alcohol laws in different countries and the number of people who suffer from alcohol dependency attempts to reject the null hypothesis. The null hypothesis states that there is no relationship between drinking age requirements and alcoholism. The researcher also predicts that countries with more stringent drinking ages will have fewer alcoholics. The following statements sum up the two hypotheses:

> H_0: There is no relationship between laws of drinking age and the number of people with alcohol problems. *(null hypothesis)*
>
> H_1: There is a relationship between laws of drinking age and the number of people with alcohol problems. In countries with stricter laws of drinking age, fewer people will drink alcohol and fewer will suffer from alcoholism. *(alternative hypothesis)*

The researcher collects data from different countries where the minimum age for drinking is nonexistent, and the data include 16-year-olds, 18-year-olds, and 21-year-olds. The study reveals that neither of the hypotheses is supported. In fact, the opposite hypothesis seems to be the case. In countries where there are lenient laws governing drinking ages or no laws at all, fewer people have problems with alcohol than in countries where these laws are much stricter. Interestingly, this researcher rejected the null hypothesis and could not support the alternative hypothesis. This researcher stumbled on something that was not predicted at the beginning of the study, nor was it supported by the literature.

Failing to Reject the Null Hypothesis

Regardless of whether we find support for the alternative hypothesis, the outcome of rejecting the null hypothesis is always desirable. But what happens when we fail to reject the null hypothesis? We still built new knowledge, which is important to recognize. Failing to reject the null hypothesis also contributes to the advancement of science.

Researchers rarely overlook this fact. However, there are those cases when a researcher is not convinced and continues with vigorous research efforts, looking for another opportunity to reject the null. It is important that scientists forgo their personal investment in the research and learn to be gratified with the fact that our knowledge is improved even when one fails to reject the null hypothesis. When we fail to reject the null, we are sometimes able to identify new relationships between variables or discover other associations we could not have imagined. New findings and new relationships are always possible. It is important to avoid discouragement because as long as studies are properly conducted, failing to reject the null hypothesis is just as great as rejecting it.

RESEARCH IN ACTION 13.1

REPORTING RESULTS

Source: Sharafkhani, N., Khorsandi, M., Shamsi, M., Ranjbaran M. Low Back Pain Preventive Behaviors Among Nurses Based on the Health Belief Model Constructs. *SAGE Open* 4(4). First published date: December-10-2014 http://journals.sagepub.com/doi/abs/10.1177/2158244014556726.

RESULTS

In this study, the mean age of the participants was 32.1 ± 5.3. Moreover, 8.3% of the participants were male and 91.7% were female with a mean work experience of 5.38 ± 3.6 years. Other demographic characteristics are presented in Table 1.

The results of the study revealed that the mean score of the nurse performance was 1.48 ± 0.27 out of 3. Performance means the proper posture and observing ergonomic principles at work, including refraining from turning around the waist axis during patient transfer or lifting loads, refraining from bending on the load (or patient), refraining from holding the arms higher than shoulders for some minutes, observing the principles of patient transfer and lifting loads (including lifting instruments, trolleys, sinks, buckets, and baskets), refraining from keeping a long distance between the patient and the nurse when moving a patient, placing the instruments on a place not higher than the waist level, using a ladder or stool to get access to objects on high shelves, and refraining from lifting heavy loads (e.g., moving sedentary patients). The mean score of nurses in observing the principles of sitting and standing was 1.16 ± 0.41 out of 3. The principles included refraining from excessive bending forward or backward, refraining from turning around, and observing the

As you can see here, the researchers are reporting the most important demographics first (age, gender, and work experience). Then, they provide a table that summarizes the rest of the results. Note how the age of participants is shown as 32.1±5.3, meaning the average age is 32 years old with a standard deviation of 5.3 years. This means that around 68% of the participants in the study are between the ages of 27 and 37 years. Also note how gender is given in percentages because it is important to see that the majority of the sample (almost 92%) is women.

(Continued)

(Continued)

correct principles regarding sitting on the chair. The nurses' mean score in observing managerial principles for adopting low back pain preventive behaviors was 1.66 ± 0.35 out of 3. The principles included reducing the frequency of patients' transfer and lifting objects to less than 20 times during a work shift, asking for help when moving patients who are unable to keep their balance, refraining from performing activities beyond one's ability, and regarding work time limitations (not working in short interval work shifts).

Finally, the nurses' performance score in performing exercises to strengthen back muscles was equal to 0.71 ± 0.55, which was too low; 33.8% of the nurses did not exercise at all, 60.9% seldom exercised, and only 5.3% exercised most of the time. Other findings of the HBM constructs are shown in Table 2.

According to Table 3, there was no correlation between perceived susceptibility and severity with the performance and between perceived benefits and performance. However, there was a reverse correlation between self-efficacy and performance, a positive correlation between cues to action and performance, and a positive correlation between the cues to action and self-efficacy.

The main barriers that the nurses confronted in adopting preventive low back pain behaviors were reported as lack of access to proper hospital equipment, inappropriate workplace conditions, high work load, excessive fatigue, and lack of knowledge on chronic low back pain prevention principles.

Next, the researchers show more advanced findings, though still descriptive, regarding the performance of nurses at work. Note how after every finding, they provide an explanation of how this was measured to put the findings into perspective for the reader. We can see that nurses followed the main principles of preventive behaviors for back pain, such as heavy lifting or transporting patients, followed by principles of posture, but were less careful about the principles of sitting and standing. Most concerning of all, their reports revealed lack of exercising to strengthen back muscles.

Finally, the researchers report on the relationship between variables. Note how this is a good example of how the researchers did not find a relationship they expected in the study, but they found another finding that was not initially expected or predicted.

The example in Research in Action 13.1 is from a study conducted by Sharafkhani, Khorsandi, Shamsi, and Ranjbaran (2014), titled "Low Back Pain Preventive Behaviors Among Nurses Based on the Health Belief Model Constructs." The researchers focused on the musculoskeletal disorders caused by the nursing profession, specifically low back pain. They employed a health belief model to investigate the relationship between nurses' educational needs about low back pain and their adoption of preventive behaviors. They collected data from 133 nurses from three public hospitals using a questionnaire about knowledge of low back pain and adopting preventive behaviors. Research in Action 13.1 provides the results from this article.

RESULTS IN QUALITATIVE STUDIES

Results in qualitative research follow a similar format to those in quantitative studies. At the very beginning of the section, the researcher provides an overview of the demographics. Providing the reader with sufficient detail regarding who participated in the study and the related demographics is essential regardless of the type of research conducted. The researcher then points out the patterns of similarities and differences that were recognized during the analysis.

Let us imagine that a researcher recognizes five similar patterns and three different attitudes or opinions on the same topic. The most straightforward way to report these results is to rank them in order of importance and then combine results where patterns exist. For example, a researcher finds out from an analysis of social media games that (1) people who play these games create a new virtual world that follows a set of rules and regulations similar to those found in real life, (2) people who invest a lot of time in these virtual worlds also create a definition of the *other* or the outcast, and (3) the attitude toward these outcasts in the virtual world are similar to social attitudes in the real world.

As you can see from these three findings, the most important discovery is that people who play virtual games are likely to create a social world that obeys rules and regulations just as in the real world. The second and the third findings about how we define the *other* and how we behave in the presence of people considered as outcasts—although insightful—are secondary points to the first.

Qualitative studies add depth to the research by providing rich and artistic detail. Using quotations in the results section is not uncommon and almost expected. These straight quotes from the text give the reader an opportunity to engage in the data and to gain a deeper understanding based on the power of the participants' comments.

RESEARCH IN ACTION 13.2

ORGANIZING FINDINGS

Source: Pajo, B., & Cohen, D. (2012). The problem with ADHD: Researchers' constructions and parents' accounts. *International Journal of Childhood Disorders, 45*(5),11–33. With kind permission from Springer Science+Business Media.

Among 12 reports with relevant information on this theme, parents' views on ADHD varied widely. In four studies, parents explain ADHD as an internally caused biological condition uncontrollable by the child (Harborne et al., 2004; Johnston & Freeman, 1997; Klasen & Goodman, 2000; Taylor et al., 2006). Some parents report persisting difficulties to make sense of the condition (Kendall, 1998) because their child's behavior appears highly inconsistent (Arcia et al., 2004). Other parents ascribe ADHD to temperament (Arcia et al., 2004) or poor parenting (Ghanizadeh, 2007), refusing to consider it a bona fide illness or disorder (Wilcox et al., 2007). Singh (2003), who interviewed mothers and fathers separately, reports large discrepancies between their views, with fathers less willing to ascribe a medical cause to their sons' behaviors. Compared with White American parents, African American, Latino, and Iranian parents put less faith in the medical diagnosis or in ADHD as a distinct condition. African Americans and Iranian parents see ADHD as resulting from a lack of parental discipline (Dennis et al., 2008; Ghanizadeh, 2007; Olanyian et al., 2007), whereas Latinos view ADHD-like behavior as normal (Arcia et al., 2004). Latinos are also concerned about the stigma of mental illness for their ADHD-diagnosed child (Fernandez & Arcia, 2004; Olanyian et al., 2007).

The focus of this study is to understand more about attention deficit hyperactivity disorder (ADHD) and the prevalence of this diagnosis among children. The researchers collected all the qualitative research conducted on parents of children diagnosed with ADHD to better understand their concerns and family dynamics. More specifically, the researchers investigated how ADHD was viewed by parents, what were the complex circumstances that pushed them toward seeking a diagnosis, and deciding to medicate their child.

You can see the logical way of presenting these findings. First a statement is made: "parents' views on ADHD varied widely." Then, the variation of this statement is explained using a spectrum, with descriptions of children's behaviors being a "bona fide illness," "lack of parental discipline," or "normal." Each variation is supported with further details and information from the studies examined.

It may be helpful to first create an outline for the results section. Ask questions such as these: (1) What patterns need to be reported first? (2) In what order should the patterns be reported so that they make the best sense to the reader? (3) How can I assemble the findings in a way that helps the reader reach the conclusion I am trying to convey? (4) How should patterns be presented such that differences or similarities are detected? Once you begin pondering these questions, you will be able to give the reader a good sense of the outcomes you wish to convey. You may choose to omit some findings and emphasize others. Some you will want to portray visually. A logical and well-organized presentation of your results will communicate your key points.

Once the patterns are presented, the researcher may opt to use diagrams to show the relationships found in the data. These visual tools can be quite powerful and are an effective means of outlining the researcher's logic. In the case of qualitative research, diagrams or figures are especially useful as they can help to lessen the pitfalls of subjective interpretation. A precise and well-rendered diagram or figure can efficiently convey information about a study from the researcher to the reader.

Figure 13.1 is taken from Sara Roncaglia's (2009) book *Feeding the City: Work and Food Culture of the Mumbai Dabbawalas.* The book is an ethnographic work of dabbawalas—workers who deliver homemade lunch-dabbas to people in Mumbai at their place of work or study. It is a delivery system that allows people to eat warm, homemade food every day. The figure represents a culinary triangle between the customers, the homemade food delivered by dabbawalas, and the influence of what the author calls *the rotten city of Mumbai.* The idea cultivated here is that people receive a cultural transformation through these home-cooked lunch deliveries and the natural process of transformation of the city.

VISUALLY PRESENTING RESULTS

Visually depicting your findings is like painting on a blank canvas. However, this is not an optional artistic endeavor, but rather expected—and often required—in presenting your research. Visual representations help to convey meaning, show dynamics, depict details, and ultimately illuminate the key relationships researchers and readers should be focusing on. Visual tools can be attractive and grab our attention immediately because our brains can perceive the entire landscape at a glance rather than deciphering data on an item-by-item basis. The most widely used types of visual presentations are tables and figures.

FIGURE 13.1 ■ The Culinary Triangle From *Feeding the City*

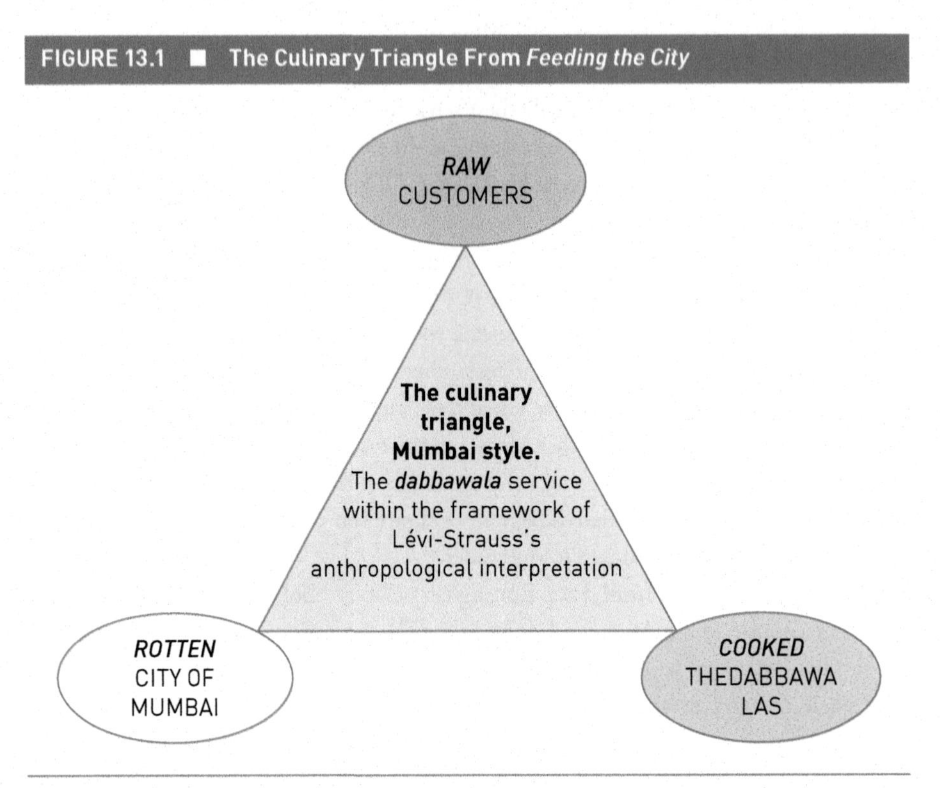

Source: Roncaglia, S. *Feeding the City: Work and Food Culture of the Mumbai Dabbawalas*. Cambridge, UK: Open Book Publishers, 2013. http://dx.doi.org/10.11647/OBP.0031. http://www.openbookpublishers.com/product/87. CC BY 4.0. https://creativecommons.org/licenses/by/4.0/

Tables

Tables can deliver a large amount of information at one time. They work best when the information is extensive and providing it in narrative form can be inefficient. Demographics are often presented in a table, for example, when the researcher wishes to avoid a review of numbers for a lengthy list of characteristics. A good rule of thumb is to present the information in a table when it occupies less space than a written rendition of the same information (Durbin, 2004). Tables are also very handy when comparing various characteristics between or across categories. For example, if we find that women are more likely to listen to music when exercising as compared to men, conveying the information in a table may be most effective.

Tables can also show a missing value. When investigating research conducted on parents of ADHD children, I remember pondering the definition of ADHD and how much it varied from one researcher to another. I collected all the information and presented it in a table, which had a decisive impact in showing that some researchers agreed with calling it a "disorder" narrative, others recognized ambiguity in the definition, and still others completely avoid defining it. Presenting these results in a table was more effective than expressing it in narrative text. Table 13.1 illustrates a part of this table.

Tables are useful and effective in conveying information, but it may be advisable to refrain from repeating the outcomes in a table and in the narrative. You can discuss the results and their implications, but you should not repeat the points you've made visually.

TABLE 13.1 ■ Table From "The Problem With ADHD: Researchers' Constructions and Parents Accounts"

Authors (Year)	Acknowledgment of ADHD Controversy[a]	Theory	Problem Statement	Recommended Application[b]
Singh (2004)	4	Y	Mother-blame for ADHD is ubiquitous; mothers might stand to gain most from the absolution promised by brain-blame	
Singh (2005)	4	Y	Bioethical analysis of issues raised by neurocognitive enhancement such as Ritalin use is detached from real-life decision-making	(Because the shift to long-acting stimulants reduces the number of moral decisions that parents must make, the resulting incremental changes in society require close and proactive scrutiny)
Taylor et al. (2006)	4	Y	In light of the ADHD controversy, how do parents decide whether or not to medicate their diagnosed child?	Prescribe long-acting stimulants to lesson stigma on children, and provide multi-modal treatment for children and resources to parents confronted with the decision to medicate

(Continued)

TABLE 13.1 ■ (Continued)

Authors (Year)	Acknowledgment of ADHD Controversy[a]	Theory	Problem Statement	Recommended Application[b]
Studies that define ADHD as a questionable entity				
Carpenter and Austin (2007)	2	Y	Regardless of what they do, mothers of ADHD children will fail to live up to the motherhood myth and will be disabled as a result	
Cohen (2006)	4	Y	Children who manifest specific behaviors in school setting are medicated, but it remains unclear how this option arises	(In complex systems of care, implicit functions override individual actors' explicit intentions; i.e., in some school settings, use of medication is a foregone conclusion for referred children even before they are evaluated)
Malacrida (2001)	4	Y	ADHD's ambiguity helps to study resistance to professional surveillance and stigmatization	(Teachers are filling an uncomfortable role in the medicalization of ADHD)
Malacrida (2004)	4	Y	ADHD's controversy complicates the routine work of medicalization that non-medical personnel carry out	

Source: Pajo, B., & Cohen, D. The problem with ADHD: Researchers' constructions and parents' accounts, *International Journal of Early Childhood* (2012), 45(5): 11–33. With kind permission from Springer Science+Business Media.

[a] Reports were rated as follows: 0, no mention of a controversy; 1, brief, non-specific, or accidental (e.g., repetition of a finding) mention of a controversy in the conclusion of the report; 2, brief, non-specific mention of a controversy in the introduction of the report; 3, specific or detailed mention of a debate over the *causes* of ADHD, itself considered a valid entity; 4, specific or detailed mention of the controversy over the *existence* or *validity* of ADHD as a clinical entity.

[b] Statements in parentheses are conclusions, not recommendations.

When you watch a movie, you would not be entertained if the director showed you an event and then told you what he or she just showed. Just as this detracts from a movie, it detracts from your research presentation. Pick the one best way to present your information and then commit to it.

RESEARCH WORKSHOP 13.1

ORGANIZATION OF TABLES

The American Psychological Association (APA) has created a set of rules and regulations pertaining to writing, which is widely used in the social sciences. In addition to writing regulations, they formulate some rules on table formatting. Some of these rules are important for a neat representation of findings:

- Tables are used to present the reader with information that would have been too repetitive and wordy if included in the narrative.
- The numbers or words need to be clear and consistent if used more than once (i.e., if the term *Caucasian* is chosen in one column/row and *White* in another, it may confuse the reader, so consistency is necessary).
- Table clarity is provided by presenting all the information in a simple way and by explaining every detail, such as symbols or abbreviations.
- The table title must be concise and accurately speak to what the table is illustrating.
- Data in a table will commonly have only two or three columns and rows need to be presented in text. If it is necessary to have additional columns for more complex data, a tabular format may be more appropriate.

For more information please read the APA guidelines (*Publication Manual of the American Psychological Association, 6th edition).*

Figures

Some common types of figures are charts, graphs, plots, and path diagrams. These types of figures are explained extensively in Chapter 10, with the exception of **path diagrams**. Path diagrams are a basic model that depicts the relationship between variables by using arrows or other types of connectors. They are often used to present theoretical background, hypotheses, or other models that either predict or present a relationship between variables and/or a theoretical concept. In addition, diagrams portray the strength of the relationship between variables by providing a snapshot of all the variables at the same time, helping to put everything into perspective.

Regardless of the type of figure you are using in your research report, there are certain guiding rules that may be helpful. First, it is important to ask questions, such as these: (1) Why am I selecting this type of figure? (2) Is this graph/chart/plot/diagram useful in illustrating my point? (3) Is this figure more effective than an explanation using words? (4) Is the figure a good representation of the findings?

Consider the path diagram in Figure 13.2, taken from an article by Lee and Randolph (2015) on effects of parental monitoring on youth. This is a cross-national study that focuses on parental monitoring in the United States and South Korea. Simply by looking at it, even if we do not understand any of the statistics reported in the diagram, we are instantly aware that lack of parental monitoring is related to smoking cigarettes and drinking alcohol (the sign is negative and there are three asterisks next to the number). We also see that parental monitoring has a positive relationship with self-esteem (the sign is positive and three asterisks follow the numbers reported). We also note that lack of parental monitoring is associated with aggressive behaviors (although the relationship seems weaker than the first ones because of the number starting with 0 and having only one asterisks). In turn, cigarette use and alcohol use are both positively related to aggressive behaviors (relationships seem stronger here) and negatively associated with self-esteem. The path diagram is also telling us that all these variables were tested simultaneously, since they are portrayed in the same path diagram.

FIGURE 13.2 Path Diagram

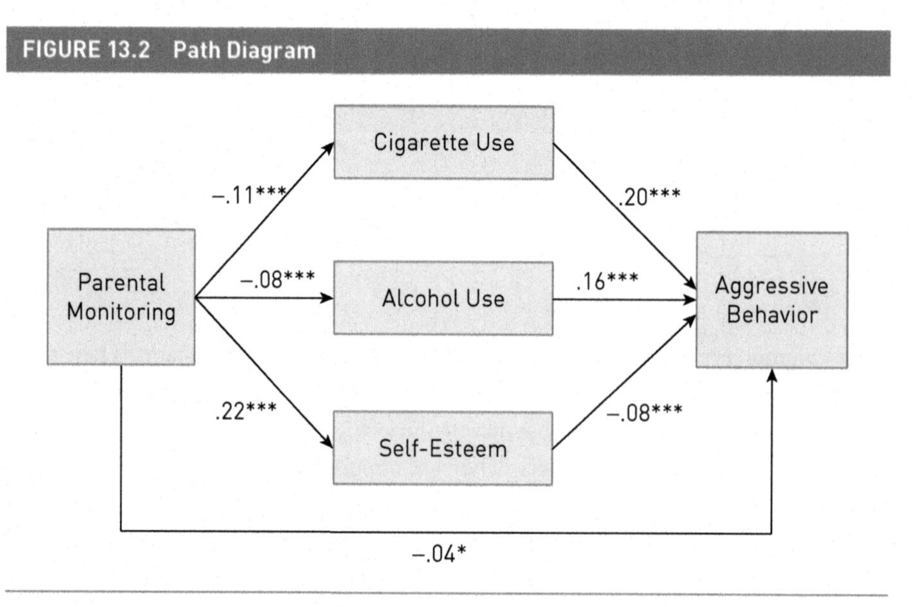

Source: Lee & Randolph (2015).

Once you are certain about every type of figure in your paper and its purpose, it is helpful to consider some other stylistic guidelines, such as labels, colors, amount of information, and clarity. All the figures should be labeled. Having a good numbering or labeling system will keep your figures in order and ensure the reader can reference them. This same simple rule applies to the use of color. If there are two or more groups represented in the figure and you are trying to make a distinction between them, it helps if they are color-coded. Obviously, it is not desirable to confuse our readers, so the best practical policy would be to use two very distinct colors. If you are publishing in a one-color format, with black- and-white being your only option, you can certainly differentiate using shades of gray or patterns.

ETHICAL CONSIDERATION 13.1

MISREPRESENTING RESULTS

It may feel like ethical issues have no place when we are presenting our results in such a straightforward way, but their presence is perhaps more important than ever. It is in the results section that the data may be misrepresented or manipulated in order to be attractive and worthy of publication. We may feel like we need some sensational results to get attention and that could have happened if only this variable was behaving slightly differently. . . . These thoughts lead to a dangerous path.

Although it may feel like you are not violating the rights of any person by tweaking the findings just a little, that is far from the truth. By misrepresenting the results, you are drawing inaccurate relationships between constructs that may create misguidance for the scientific community. In other words, you are violating the rights of an entire community—not just one specific person. Sometimes people may feel obligated to exaggerate their findings to receive grant money or to justify the grant money they have already received. Regardless of the circumstances, exaggerating, misrepresenting, tweaking (such as ignoring missing data or other seemingly harmless changes), or any other intentional modification of the results is a form of deception and highly unethical for the entire scientific community. Your own findings—no matter how small—are accurate, important, and the best representation of your dignity and work.

DISCUSSION

The discussion is perhaps the most creative part of any research report. It weaves together everything from theoretical background to the results to the interpretation of the results. In a nutshell, the discussion is where you write about your interpretation of the results and what comes next. It is also the section that makes or breaks your write-up. It shows how much the researcher has invested, has carefully considered the findings, and the time and energy dedicated to the study. Basically it answers the question: What do you make of all of it? Writing a successful discussion is easier said than done.

Your discussion should begin with a clear outline of what will be included and how. Usually the researcher delineates the limitations of the study before beginning the discussion, but there is wide variation in this method. Some articles will include the limitations as a section on its own right before the discussion even begins, while others use limitations as a way to get the discussion going.

Regardless of personal preference, it is helpful to address the following question: What could have increased the validity and reliability of the study? As you know from Chapter 5, there are different types of reliability, all referring to consistency of measurement. Validity applies to the accuracy of the measurement. Limitations explore the factors that could have improved the effectiveness, such as a bigger sample for more reliable results or a more diverse sample. Perhaps one of your questions went awry during data collection and you are prevented from including an entire variable in the analysis. This is the time to mention the imperfections of the research.

The logic of addressing the limitations before the creative discussion is almost like a disclaimer. It says, "We concede there are imperfections in this study before we discuss how the findings relate to the new knowledge we are building." The findings are then accepted at face validity, regardless of any imperfections.

Writing a solid discussion section takes good organizational skills. If the literature prompted to hypothesize that college athletes are less likely to engage in risky behaviors, such as alcohol and drug use, but our findings reveal the opposite, we may consider restating our hypothesis in the discussion. In one simple sentence, we can represent the rationale for our hypothesis and connect it to our findings.

Furthermore, we are expected to interpret our findings. Why did we get the results we did? One way of organizing the discussion of findings would be by order of importance. Present the most important findings first and then move on to the rest of the results.

The interpretation of results is our opinion of the findings, but it is most often based on the literature or the theories we previously researched. Our interpretation of the theories and the research is our own, but there are always other forces that influence our interpretations. If you think about it, every opinion or interpretation you have is based on your experience, background, something you have read in the past, or something you have been exposed to. Interpretations rarely come out of thin air. This may require some searching and contemplation on your part, with the intention of trying to understand why you think the way you do. This process will give meaning to your research, to the way you think, and will teach you much more than simply how to interpret findings. Research in Action 13.3 is an example of a small study that compares "how to" books targeting parents of typical children and "how to" books targeting parents of children with ADHD.

RESEARCH IN ACTION 13.3

ILLUSTRATION OF ORGANIZING THE DISCUSSION

Source: Pajo, B., & Stuart, P. (2012). Comparative review of "how to" books for parents of ADHD children and "how to" books of parents of typical children. *Children and Youth Services Review, 34*(4), 826–833.

The examination of these two sets of self-help books suggests that whether or not their authors have in mind children diagnosed or not diagnosed with ADHD, the children are portrayed as exhibiting similar behaviors in similar settings. Both groups of children are portrayed. . . . What seems to mark the distinction between a typical and an ADHD child is the perception of the frequency of such behaviors. . . .

> The discussion starts by restating the findings. The most important finding is that these behaviors of children (regardless of whether they are diagnosed or not) are similar behaviors. Children with ADHD seem to demonstrate these behaviors more often.

The tactics and strategies recommended to the parents of ADHD children focus on behavior control whereas the tactics and strategies recommended to parents of typical children focus on children's emotional wellbeing. Furthermore, the same tactics and strategies recommended to parents of ADHD children are considered as ineffective methods that perpetuate undesirable behaviors to parents of typical children.

> Further, more details from the findings are reported because the authors are focused on discussing these details further.
> A problem is identified and pointed out.

(Continued)

(Continued)

Goffman's understanding of the perception of "self" supports this idea. In his work on the social situations of mental patients, he claimed that the conceptualization of "self" is to a certain extent a result of the kind of behaviors others adopt when interacting with an individual. . . . [B]y approaching children through a disability perspective, reminding them about what they need to do, as self-help books for parents of ADHD children suggest, it is possible than ADHD-like behaviors are reinforced rather than extinguished. It is intriguing, if not disconcerting, that the self-help books targeting parents of ADHD children make no mention of how parents of non-ADHD children might be advised to approach these similar behavioral problems with their children.

Theoretical reasoning is brought into the discussion to support the argument the authors are making. Note how the path is prepared to make the case as to why suggestions made to parents of children with ADHD have the potential to make children more disabled rather than helping them.

A slight critique in the form of "wonder" is put forward here by asking why parents of ADHD-diagnosed children are left in the dark as to how parents of typical children handle these same behavioral problems.

RECOMMENDATIONS

Toward the end of the discussion, most researchers anticipate how the study might be followed up with other studies to reveal even more knowledge and insight. It is a forward-thinking process that allows the researcher to consider future options by asking, "If I were to conduct an additional study on the same topic, what would be the best approach? How would I follow up with these findings?" Putting the findings in perspective is crucial to future studies.

There are two common types of recommendations, but these are not exclusive: (1) recommendations on the methodology of future studies and (2) recommendations on topics. Considering changes in the methodology comes naturally to most scientists because the findings reveal information that could not have been predicted at the beginning. That is the beauty of conducting research. Some researchers wish they had more data about a specific characteristic, while others wish they had some qualitative information in addition to the quantitative information, or vice versa. Others might consider increasing the number of participants in the next study as the key to more reliable findings because their sample was not as diverse as they had hoped.

The same logic applies to the topic of the study. Findings may reveal the need to follow up in a specific direction that may not have been the original focus. It is often the case that findings bring an unexpected outcome that calls for further exploration. During the work, the researcher may have also changed the focus and has more of an interest in investigating another topic by the end of the study. After all, a thirst for knowledge mandates continual adjustments to the way we think, interpret, and search for answers. The beauty of conducting research reveals itself when we can change how we think and trace these changes to the trajectory of the research itself.

RESEARCH WORKSHOP 13.2

TWO WAYS OF ORGANIZING THE DISCUSSION

There are many creative ways of organizing your discussion section, but the following are two suggested outlines:

DISCUSSION OUTLINE I

- State the limitations of the study
- Remind the reader about the literature on the topic
- Restate the hypothesis
- Restate the important findings (or the findings that will be discussed)
- Start discussing and interpreting every single finding that was mentioned in the previous section, preferably in its own subsection
- Make recommendations for future research with changes in methodology and topic

DISCUSSION OUTLINE II

- The limitations appear before the discussion
- Start the discussion by restating the findings that will be discussed
- Discuss each finding in its own subsection
- In each subsection, start by recalling the literature that led to a specific hypothesis
- Restate the findings and interpret them by linking them back to the literature
- Make recommendations specific to this finding
- Move on to the next subsection, and begin the process over again

Summary

Reporting the results of your study is one of the simplest sections of the research report to create. The researcher begins by restating the hypothesis and describes to the reader where the idea for the study originated. The next step is to provide a good description of the demographics and frequency of each variable. The language used in this section is straightforward, so the reader can clearly understand the findings. The researcher reports on the relationship between variables of interest and states whether there were any other findings besides those hypothesized at the beginning of the study. We can reject or fail to reject the null hypothesis. Results are reported in both cases because we advance science regardless of whether we were able to support our hypothesis.

When reporting results in qualitative studies, we follow rules similar to those for quantitative studies. First, the researcher restates the focus of the study. Initially, some demographics or an overview of the participant pool is given before moving on to the findings. In qualitative studies, the researcher may incorporate direct quotes to illustrate the findings. Often, the results section for qualitative studies reports on patterns of similarities and differences that were found in the study.

Visualization of data is used to illustrate information from the findings that adds to what is written in the text. Sometimes the details are too numerous to be expressed in text, so the researcher writes about the most important findings and includes the other details in a table or a figure. Figures or tables should not repeat information presented in the text and should instead add new information and insights. Finding the best type of visualization for the study will make the report more engaging for readers. While tables, charts, and graphs are often used to depict frequencies or relationships between variables, path diagrams are used to visually represent the entire model of the study. Path diagrams help to put all the variables in perspective and make it an easy, clear way for readers to understand the study's process.

The discussion follows the results section. In this section, the researcher is able to draw interpretations from the findings, explain the results, and connect them to the theoretical framework used at the beginning of the study. Discussions can be organized in various ways, and a logical organization of the discussion can turn a good research report into an outstanding one. Prior to the discussion, the researcher may discuss the limitations of the study, acknowledging potential problems for interpreting the results. Finally, the discussion section ends with recommendations for future research. What type of study would be a good follow-up, based on these results? How can we advance science further? These recommendations are crucial for the reader who is interested in researching the topic further.

Key Term

Path diagram: a model that depicts the relationship between variables by using arrows or other connectors.

Taking a Step Further

1. Why is the representation of results the simplest section in a research report?
2. What are some ground rules to keep in mind when reporting results?
3. What type of information is best to visualize rather than report in lengthy writing?
4. What is the purpose of the discussion section?
5. What type of connections with other sections of the research report do we draw in the discussion section? Why is this section considered to be creative?
6. Why is failing to reject the null hypothesis as important as rejecting the null hypothesis?

PRESENTING YOUR RESEARCH

CHAPTER OUTLINE

WHAT WILL YOU LEARN TO DO?

1. Explain how best to present your study to an audience
2. Discuss how to apply to regional and national conferences
3. Describe how to get an article published in a peer-reviewed journal

PRESENTING YOUR FINDINGS TO AN AUDIENCE

You have done everything you can to conduct the best research study possible, but something is amiss if no one is aware of what you discovered. Presenting your study is perhaps as important as conducting it, because this is how most people will be able to understand your procedures, discuss results, offer feedback, and take your work a step further. You are probably familiar with presenting in front of your classmates by now, but your presentations have likely focused on presenting others' ideas or research results. When it comes to your own studies, you will have the exciting opportunity to tell the world about what you did, how you did it, and the contribution you made.

© iStockphoto.com/Rawpixel

Presenting our findings to an audience is our chance to bring attention to the importance of our findings. It is our tool to communicate our study to the world.

Your classmates are the first audience for your work. Although you may feel comfortable with them, you should still take their presence seriously. Envision your friendly classmates as unfamiliar academics who will ask questions, evaluate your research, and provide feedback. They are not familiar with your topic, nor have they read your report. Since they have no idea how you conducted your research, your presentation is your opportunity to describe this process in a 10- to 15-minute time frame.

Presentations are crucial in spreading the word about your work and, if done well, can open the door to additional opportunities, such as presenting at regional or national conferences, publishing a journal article, or even pursuing related topics in your graduate work. For the moment, let us return to the topic of preparing the best presentation possible in the time allotted. How do you convert your potentially uninterested audience to one that is fully invested? How can you involve and engage them, encourage questions, and even inspire them? Since you have limited time, the details need to be thoroughly thought out.

Identify the Main Points

When you are presenting your research, think about the entire project first rather than immediately creating PowerPoint slides. Consider what this project is really about.

You will provide information about its background, the literature review, the methodology, results, and discussion points, but which of these are most important to focus on? In order to attract the audience's attention, which of these areas deserve your time and focus? Your findings may be most important, but you may also want to focus on data collection or an aspect of the analysis. Whatever your main ingredients, identify and record them. This will help clarify your central points before you begin planning your presentation.

Rank Your Topics

Three to four points will define your research. Consider how much time you want to spend on each aspect. Is one point more important than the others and therefore warrants more time? Remember that you will need to provide background information to put the study and its concepts in context. However, do not provide too many details of the literature review when you should be focusing on the study's findings. Otherwise, your audience will wait and wait and suddenly someone will start snoring in the back of the room. You must know how to manage your time so you can quickly move through your presentation to the most exciting part of the study—the part that makes you feel interested and alive. In other words, rank your key points so that you know in advance exactly where you'll spend your time.

Tell a Story

From an early age, we are fascinated by storytelling. Our attention is naturally captured by the details of any good story. If you give people a list of statistics, they may nod and smile, but they will never be as engrossed and involved as when they hear a human-interest story. Put this tip to good use and incorporate a relevant story that captures the essence of your study. This works wonders at the beginning of the presentations, but it will grab the attention of the audience at any time.

The first time I ever presented my work, I assumed I would appear credible by opening with statistics. "Ten percent of U.S. children under 15 years old take psychiatric medication to control their behavior!" I claimed. I encountered blank stares and could sense the audience's indifference. "So what?" they were thinking. "We all take medication in one form or another, and 10% is not that high anyway." I was puzzled as to why the audience did not seem to care about the numbers I was presenting, but I should not have been. Numbers are truly powerful, but they are not emotional and do not touch us as powerfully as true stories.

The next time I had a chance to address a large audience, I did not repeat this mistake. I told them about a 4-year-old girl who passed away while on Ritalin, a methylphenidate prescribed to children to control their attention deficit hyperactivity disorder (ADHD)-like behaviors. A brief 20-second video of the 911 call of the mother about her daughter's sudden death made all the people in the room freeze. The audience immediately responded with compassion, sadness, concern, and, most importantly, close attention to the importance of my message. I had proven that the psychiatric medications we give to children may be really dangerous. Now the statistic of U.S. children who take medications had a much more powerful effect than in my previous presentations.

I had a valid topic of study, and the audience was eager to know more about how I conducted my research. Some topics lend more drama than others, but I am certain you can find your own story to tell. You were probably attracted to your research because of a personal story or connection to the topic. You should tell that story or connection; whatever it is, you should illustrate it so that everyone who is listening can somehow relate to your story and become drawn to your topic and presentation.

ETHICAL CONSIDERATION 14.1

ACCURATE PRESENTATIONS AND ANONYMITY

Although we want the audience to gain interest in our study and sympathize with our cause, we also want to be careful about the ethical standards of research. It is our responsibility to accurately present data and not mislead the audience on our findings or study details. Clearly, we want to present our work in the best light possible and we are not violating any ethical standards if we present the same information in the most attractive way. We simply need to be extra careful to not misrepresent the information or depict an inaccurate picture of events.

Furthermore, we need to make sure that we are not harming anyone or violating anyone's privacy through our presentation. The death of the 4-year-old girl using Ritalin was widely publicized in the media and was also featured in the PBS movie *Medicating Kids*. If this case had been kept away from social media because of her parents' personal wishes, it would have been a violation of their rights to make it public through a presentation. In other words, we cannot find and use personal cases that may work to draw attention from an audience, but will violate their anonymity or any other ethical agreements.

Literature, Hypotheses, Methodology, and Findings

Now that you have the audience's attention, you need to show them that you are a professional. While your story is attention-grabbing, you also need to establish that your work is professional, ethical, and thoughtful. Keep it brief, but mention what the literature offers. What do other studies say about the topic? What were some of the gaps in the literature? How did the literature drive your hypotheses and study?

Briefly state your hypotheses and the relationship you were trying to investigate. Then, be succinct about your methodology, starting with ethical considerations and safeguards to protect the anonymity of participants. Discuss how the data were collected, the number of participants, and participant demographics, including race and ethnicity, gender, religiosity, sexual orientation, and single-parent versus two-parent families.

It naturally flows if you address the findings following the discussion on data collection. You will revisit the hypotheses and state whether you rejected or failed to reject the null hypothesis. Did you find something more about the relationship between the variables of interest? Was there something you missed? Did you discover something new? What were the limitations of your study? Be as professional as possible. This is the time for specific and factual descriptions so the audience will know exactly how your study was conducted.

Visual Aids

It is not by chance that this discussion of visual aids comes at the end of the section. Visual aids are not the focus of your presentation. They are your aids and will help convey what you want the audience to learn from your study. They are not a goal in themselves, although they are powerful attention grabbers. Whether you use a short video clip, an animation, slides, or another kind of visual representation, it should fit the needs of your study. It is typically helpful to use a visual aid, but remember that what you are saying should be the focus of the presentation. If you have text on your slides, keep it to a minimum. Fewer words equals a better presentation. You would be wise to show tables, graphs, figures, or other visual illustrations of your findings rather than any words at all.

Practical Tips

The tone of your voice and your body language when you address the audience are important in making an effective presentation. When you are in front of a group of people, everyone is looking at you. It is necessary to make eye contact with them. It is not only polite, but makes for a more effective presentation. Have you ever met someone who does not look you in the eye when you talk to them? It can make you feel uncomfortable and a bit ill at ease, right? If you start looking people in the eye when you talk, it will build confidence and increase your ability to incorporate humor, tell stories, convey your research, and otherwise have a great time. If you avoid looking at your audience, you will transmit your insecurities, make everyone feel uncomfortable, and quickly lose their attention. It goes without saying that if you read your slides verbatim you are almost saying, "I have no desire to look at you, so I am reading these to you."

Another important rule is to keep your voice in check. A calm yet strong voice, pausing as you speak, allows for two crucial things to occur: (1) it will show you are confident about your study, and (2) it will allow you to formulate the next sentence before you speak. Rushing your presentation makes the audience jittery and inattentive; going too slow may cause them to lose track of your point. Practicing a good speed and tone of voice is the key to finding a perfect rhythm.

Humor can be engaging in a presentation, but forced humor can be disastrous, so use caution and think about whether it is right thing for your presentation or not. Dry humor or sarcasm usually works best. Be comfortable with who you are as a person before deciding on whether you would use humor in your presentation. Stand-up comedians direct a good amount of their jokes at themselves, which may work well for you too. Humor is delicate, so unless you have experience with it and feel confident using it, it needs to be treated carefully.

Finally, dress appropriately and stand up straight. Your audience wishes to feel appreciated and putting effort into your appearance is one way of saying, "I am showing you respect." Slouching, hiding behind a podium, or touching your face or hair shows fear and insecurity. You must present the best version of yourself and if it feels uncomfortable for the first few seconds, you will see how quickly you will get accustomed to it. Some presenters are good at using their hands. Learn how to use your hands skillfully, and you will have an engaging presentation. Think about it. Dramatic people are the ones we listen to the most because they do not simply talk with their mouths, but with their entire bodies.

There is much information available on the topic of presenting. If you want to learn more about how to become an excellent presenter, see *The Art of Public Speaking* by Stephen Lucas; *Resonate: Present Visual Stories That Transform Audiences* by Nancy Duarte; and *The Naked Presenter: Delivering Powerful Presentations With or Without Slides* by Garr Reynolds.

APPLYING TO CONFERENCES

When you started your first research study, you may not have considered the outcome or what to do with your findings once the work was completed. Presenting your work to your classmates is a fulfilling endeavor and helps you practice this skill. You should strive, however, to present your findings at regional or national conferences in your field. Many organizations and universities organize their own annual conferences, but it is perhaps advisable to consider applying to a smaller conference first.

Local universities will likely host conferences open to undergraduate students. Start browsing to see whether your topic fits within the focus of one such conference. Some professors create student email lists to publicize articles, conferences, local jobs, and other events. Approach your professors and ask whether you might be included in their lists or whether they have information about regional conferences where you might present. Conferences that target undergraduate research are the most beneficial because you can present your research in front of an audience, listen to feedback, get used to answering questions, and network.

Regional conferences help to hone your presentation skills and prepare you for the next step: national conferences. A national conference is more competitive and usually accepts only poster presentations from undergraduate students. You may be thinking that presenting a poster isn't fulfilling, but, in fact, becoming familiar with researchers in your field, participating at the conference, talking to professors from various universities, and listening to other research presentations are all extremely beneficial. You may be able to add the poster presentation into your curriculum vitae.

Conference presentations help us reach out to larger audiences, strengthen our communication skills, and pave the way to networking.

Poster presentations are located in a separate area at prescheduled times and are very well attended. They afford an excellent opportunity to network with other students who are also invested in research, as well as professors from other universities. It is a great way to meet others who are interested in your topic. Students who are interested in pursuing graduate degrees should take advantage of this networking opportunity. Students interested in an academic career may also take advantage of the many opportunities and resources available at national conferences.

RESEARCH WORKSHOP 14.1

APPLYING TO CONFERENCES

Most national and regional conferences have *calls for abstracts* that often circulate among faculty or other email lists. It may be worthwhile to explore some of these conferences, look at their websites, and become a member to receive their newsletters. This will guarantee that when the call for abstracts is open, you will receive an email notification. Nationally recognized conferences may be easy to find, but the regional ones may require more digging. If you live in an area with a few

(Continued)

(Continued)

universities, they may be the first place to start looking. Many universities organize small conferences that provide undergraduate students in their school and adjacent schools an opportunity to present a paper or a poster. These are golden opportunities to get to know and network with others and to learn more about universities that you might attend for your graduate degree. To find these regional conferences, it may be helpful to talk to professors in your field of study. They will know what types of small conferences take place in the area and the approximate time when these calls come out.

Once you know the conference you are applying to, all you need to do is prepare your abstract. In most cases, you only need to submit an abstract for a paper or poster presentation. Keeping organized and submitting your abstract by the deadline is necessary, but submitting a high-quality abstract is just as important. It may be helpful to have a professor read over your abstract and give you feedback. You need to make sure that it truly represents the poster or the paper you have prepared, that it is clear, that it has an engaging style, and that it is within the word limit indicated in the call for abstracts.

Once you submit the abstract, it goes to a committee of conference organizers, who read these abstracts closely, discuss their importance, and approve or reject them. National conferences are more competitive because more people apply to them, so their deciding committees are inundated with interesting and engaging abstracts. If you apply to a national conference and are rejected, you should not become discouraged, but try again the next year. The chances of being approved are slightly better for regional or university conferences, and they can be a good starting point if you have never applied to one before.

PUBLISHING AN ARTICLE

One of the goals of research is to make the findings available to the scientific community. Presenting at conferences is one way of doing this. Another way of reaching the scientific community is to publish your study in a journal specific to your field. Many journals are open to undergraduate students who have conducted excellent research and are willing to go the extra mile in publishing their paper. Table 14.1 shows some of the common journals available to undergraduate students from different disciplines. In addition to these journals, many universities have their own research journals dedicated to undergraduate research.

Most of the journals in this table are peer-reviewed, which indicates a higher quality of article with a sound methodology. A peer review guarantees the quality of research. For example, if you are conducting research that focuses on people who have an alcohol

TABLE 14.1 ■ Common Journals Dedicated to Undergraduate Research

Discipline	Journal, Information, and Website
Multidisciplinary	*American Journal of Undergraduate Research* is a national, peer-reviewed, multidisciplinary, independent journal that publishes four times per year. Website: http://www.ajuronline.org/
Anthropology	*Anthrojournal* is an open source journal in anthropology that shows mostly scholarly work of undergraduate and graduate students. Website: http://anthrojournal.com/
Exercise Science	The *International Journal of Exercise Science* is a peer-reviewed journal that engages undergraduate and graduate students as authors and reviewers in the area of exercise science. Website: http://digitalcommons.wku.edu/ijes/
Mathematics	*Involve* is a peer-reviewed journal that aims to show cutting-edge research from undergraduate and graduate students in mathematics and related disciplines. Website: http://msp.org/involve/about/journal/about.html
Sociology, Anthropology	The *Journal of Undergraduate Ethnography* is an online international journal focused on ethnographic research studies conducted by undergraduate students. This journal publishes two issues per year. Website: http://undergraduateethnography.org/
Social Sciences	The *Journal of Integrated Social Sciences* is a peer-reviewed journal dedicated to disciplines such as political science, psychology, sociology, and gender studies. Website: http://www.jiss.org/
Multidisciplinary	The *Journal of Student Research* is an interdisciplinary peer-reviewed journal, entirely online and freely accessible. Website: http://www.jofsr.com/index.php/path
Biology	The *Journal of Young Investigators: The Undergraduate Research Journal* is a peer-reviewed journal accepting research manuscripts from various disciplines, such as biology, physics, mathematics, engineering, and psychology. Website: http://www.jyi.org/
Education	*Learning and Teaching: The International Journal of Higher Education in Social Sciences* is a peer-reviewed journal dedicated to teaching methods in higher education through social science lenses. Website: http://journals.berghahnbooks.com/ltss/index.php?pg=home

Being able to publish our study and findings is more than simply an effective way to build our resumes. It shows the strength of our study and its appeal to the research community.

dependency, you will need the opinion of experts in addiction to truly understand the quality and impact of your study. Someone who is investigating cognitive development of children would likely not have the expertise to evaluate the quality of your research.

We depend on researchers who work in related areas to critique and evaluate our work. This way, the quality of a study is evaluated prior to publication. In practice, article submissions are received by the journal's editor. Keeping in mind the journal's needs and space for publication, the editor chooses among the available articles that are the best fit and that are highly likely to succeed in the review process. Then, the article is sent to experts on the same research topic who review, critique, and make recommendations to the editor. To avoid potential biases and preferences, this process is typically anonymous. Once reviews are completed, the editor decides whether the article is suitable for publication.

There are writing standards to keep in mind when you are trying to publish your work. Start by considering what reviewers look for when they evaluate an article for publication. Hoogenboon and Manske (2012) have prepared the following list of reviewers' priorities when considering a study: (1) the relevance of the problem, (2) the writing style, (3) the study's design, (4) the quality of the literature review, and (5) the sample size. Let us take a look at each of these specific characteristics of the article.

Relevance of the Problem

Our family members may be the most intriguing, inspiring, and interesting subjects for us. We may want to conduct study after study on them. However, the world may not share our enthusiasm—they have their own family members to worry about. In other words, the focus of our study should be of interest to a broad audience. Is the study attempting to solve or offer insights on a problem that concerns our world today?

A topic that starts as a personal issue is in fact relevant to the scientific community. For example, if we were to investigate the impact of divorce on children because our own parents went through a divorce, our study will likely be relevant in today's world because of the increasing number of children in similar situations. Although it is important to find a topic that is relevant to a lot of people—a *hot* topic—it is even more important to stay true to your own areas of passion and interest because you will produce the highest quality of research when you invest your time into something you truly love to do.

Writing Style

Writing is a learned skill. The more experience and practice you have with it, the better you will become. There are many excellent guides available about approaches to scientific writing, but the following are some simple practices to improve your writing style:

1. Read as many scientific articles and books as possible to familiarize yourself with different styles, commonly used phrases and words, and writing with clarity.
2. Write every day, even for just 15 minutes. Think of writing as a muscle in your body that needs constant exercise.
3. Think ahead about the design of the study and the presentation of your work.
4. Avoid unnecessary information and words.

Study's Design

Reviewers consider the type of methodology used and whether it is appropriate for a study. As you have repeatedly read throughout this book, the design should serve the study, not the other way around. This is what reviewers will notice. Did your design selection truly fit the study or was it guided by convenience? Were you ethical in your data collection, and what measures did you take to protect your participants' confidentiality? If you conducted interviews, how did you handle the data afterward? How did you transcribe and code your interviews? If you conducted surveys, what were the steps taken to clean and code them? How many people were involved in the analysis? What is the level of sophistication you bring to this study? Having a strong methodology with a good rationale is your protective shield in the review process. There is nothing more satisfying than a study that is accepted and praised by the scientific community.

Quality of the Literature Review

Your familiarity with the work of other researchers in the field will be apparent in your literature review section. If you have scrutinized and systematically organized the literature in your area of interest, you will understand the implications. You will know not only how many studies were conducted on the topic, but the types of methodologies used,

RESEARCH IN ACTION 14.1

ILLUSTRATION OF A PRESENTATION

Source: In the eye of the beholder: Reports of autism symptoms by Anglo and Latino mothers by Blacher, J., Cohen, S. R., and Azad, G. (2014). *Research in Autism Spectrum Disorders, 8*, 1648–1656.

The following is an excellent example of a clear, simple, and straightforward presentation of an entire study. To read the actual publication, please visit the journal *Research in Autism Spectrum Disorders*, Volume 8, and look for the title of this article. The article's subheadings are presented here, which provides a good example to follow when writing an article or paper for your study. Look at how the authors have slowly and logically presented their literature review, methodology, findings, and discussion.

From the title, we can directly see what the article is about without reading it. We can see how this research is organized around the headings and subheadings before reading it. The introduction or the literature review seems to be logically structured. It first presents the cultural beliefs about childhood, then specifically discusses the prevalence of autism and how it is diagnosed among Latino children. At the end of the literature review, the researchers present us with their questions and hypotheses.

INTRODUCTION

1.1. Cultural beliefs about child development and disability

1.2. Autism diagnosis and Latino children

1.3. Research question and hypotheses

In the methods section, the researchers discuss how they measured their concepts and how participants were interviewed or surveyed. We see that the researchers revised their autism diagnostic interview, which they explain in this section. There is information on how the study was conducted and how the data were handled and analyzed.

METHODS

2.1. Measures

- 2.1.1. Intake form
- 2.1.2. Autism Diagnostic Observation Schedule
- 2.1.3. Autism Diagnostic Interview Revised

2.2. Procedure

2.3. Data analysis

RESULTS

3.1. Ethnic differences on the intake form

3.2. Ethnic differences on the ADI-R

3.3. Ethnic differences on the ADOS

3.4. ADOS classification and mother-report measures: Differences by ethnicity

Now, we move to the results section. The researchers discuss various ethnic differences among participants, ethnic differences in the autism diagnostic observation and the revised interview. These ethnic differences seem to appear on mothers' self-reports as well.

DISCUSSION

4.1. Limitations

4.2. Conclusions

The researchers then discuss their findings, describe the limitations of the study, and conclude their work. There are likely details we are missing without reading the actual research report, but the subheadings help us glean a good idea of what this work is about. This kind of elegance and simplicity is what makes any study great, which should be your ultimate goal in reporting your work.

the demographics of the populations explored, and the theoretical perspectives adopted. You will also know the gaps in the literature and can identify the missing areas in your review. A great literature review sets the tone for a great study because the reader can see you have mastered the topic and are well acquainted with the literature.

Sample Size

Having only a small number of participants can limit the impact of any study. If you surveyed only your classmates, for example, you would lack diversity in demographics, such as diversity of opinion, age, education, and socioeconomic status. Your findings may not be as informative as a larger sample that includes more participants from various backgrounds.

A larger sample size results in a study with stronger validity and reliability. Therefore, it is not surprising that reviewers consider the sample size to be one of the most important characteristics of great research. Sample size is not only important for quantitative studies. It is also a characteristic to look for in qualitative research, though it will depend on the topic of the study and how difficult it is to recruit the population of interest.

Summary

Conducting the best study possible is a researcher's aim in all circumstances, but being able to present the work to the entire world requires additional skills. Being able to present your best self to an audience is something that can be learned and improved with practice. Making your work public and presenting it to an audience is necessary in today's world. Presenting in front of your classmates is just the first step that is meant to whet your appetite for attending conferences, presenting in larger audiences, and, ultimately, publishing your work in a journal.

First, after presenting to your classmates, applying to regional conferences or conferences organized by local universities in your area may be the next logical step. It will provide you insight into the audience's response to your study and also help you network. Networking with people who are interested in the same topic opens the door for future endeavors, from attending the graduate school of your choice and getting help with future research to furthering your career in a specific area. Regional conferences can be less intimidating than national conferences, although as an undergraduate you are likely to be limited to poster presentations. However, national conferences allow for a large range of opportunities, including meeting researchers and scientists in your field (i.e., people who have written articles in your literature review) and becoming familiar with universities you may want to attend for your graduate studies.

Whether you are presenting a paper or poster, there are some best practices to keep in mind. Having a simple, easy-to-follow presentation is a great benefit. Identifying the main points of your presentation and maintaining focus on these points will give the audience a good sense of the study. You have done a great job when you interest people in your topic and they ask questions to learn more. If you are presenting a paper, rank your topics in such a way that you spend most of your time on the most interesting and engaging points of your study.

Capture the attention of your audience by either telling a story, maybe using humor, or adding a personal touch to your presentation. Meanwhile, stay professional and never forget to say something about the literature review, your hypotheses, your methodology, and the findings. While visual aids work best for large audiences, they should serve the study and not the other way around. If you are using slides, make sure that your slides have the fewest words possible and you make eye contact with the audience.

Writing an article that is suitable for publication in a scientific journal is another way to make your work available to larger audiences. There are many scientific journals available to undergraduate students, and having an article published by the time that you graduate with a bachelor's degree will put you in the *yes* pile for most graduate programs. Most of the scientific journals for undergraduate students are peer reviewed. The peer review process means that your study is read and critiqued anonymously by scientists who are experts in the topic of interest. Some common aspects of your study that most reviewers pay attention to are (1) the relevance of the problem, (2) the writing style, (3) the study's design, (4) the quality of the literature review, and (5) the sample size. Though being accepted for publication is a longer, more rigorous procedure, it ensures that published articles are of the highest quality.

Taking a Step Further

1. What are some techniques we use to capture audiences in our presentations?
2. What are some ethical considerations we need to pay attention to when presenting our study to an audience?
3. What are some best practices in using visual aids to our advantage?
4. What are some important practical tips to keep in mind when applying to a journal for publication?
5. What makes for a high quality of literature review?
6. What is the relevance of your topic and what elements do you want considered when applying to a journal for publication?

REFERENCES

Chapter 1

American Psychiatric Association. (2000). *Diagnostic and statistical manual of mental disorders* (4th ed., text revision (DSM-IV-TR). Washington, DC: Author.

Bird, A. (2013). Thomas Kuhn. In E. N. Zalta (Ed.), *The Stanford encyclopedia of philosophy.* Stanford, CA: Center for the Study of Language and Information.

Centers for Disease Control and Prevention. (2003). *Behavioral risk factor surveillance system survey data.* Atlanta, GA: Author.

Centers for Disease Control and Prevention. (2010). *Increasing prevalence of parent-reported attention deficit/hyperactivity disorder among children—United States 2003 to 2007.* Atlanta, GA: Author.

Kuhn, T. (1962). *The structure of scientific revolutions* (3rd ed.). Chicago, IL: University of Chicago Press.

Popper, K. (1959). *The logic of scientific discovery.* London, UK: Hutchinson.

Stangroom, J., & Garvey, J. (2015). *The great philosophers: Sir Karl Popper, Jean-Paul Sartre and Michel Foucault.* London, UK: Arcturus Publishing.

Thornton, S. (2014). Karl Popper. In E. N. Zalta (Ed.), *The Stanford encyclopedia of philosophy.* Stanford, CA: Center for the Study of Language and Information.

Chapter 2

Igbo, J. N., Onu, V. C., & Ohiyo, N. O. (2015). Impact of gender stereotype on secondary school students' self-concept and academic achievement. *Sage Open, 5*(1).

Mahler, S. J. (1995). *American dreaming: Immigrant life on the margins.* Princeton, NJ: Princeton University Press.

Pajo, E. (2007). *International migration, social demotion, and imagined advancement: An ethnography of socioglobal mobility.* New York, NY: Springer.

Chapter 3

Arcia, E., Fernandez, M. C., & Jáquez, M. (2004). Latina mothers' stances on stimulant medication: Complexity, conflict, and compromise. *Journal of Developmental and Behavioral Pediatrics, 25*(5), 311–317.

Blacher, J., Cohen, S. R., & Azad, G. (2014). In the eye of the beholder: Reports of autism symptoms by Anglo and Latino mothers. *Research in Autism Spectrum Disorders, 8,* 1648–1656.

Bussing, R., & Gary, F. A. (2001). Practice guidelines and parental ADHD treatment evaluations: Friends or foes? *Harvard Review of Psychiatry, 9*(5), 223–233.

Cohen, D. (2006). How does the decision to medicate children arise in cases of "ADHD"? In D. Cohen, G. Lloyd, & J. Stead (Eds.), *Critical new perspectives on ADHD* (pp. 236–252). London, UK: Routledge.

Dennis, T., Davis, M., Johnson, U., Brooks, H., & Humbl, J. (2008). Attention deficit hyperactivity disorder: Parents' and professionals' perceptions. *Community Practitioner, 81*(3), 24–28.

DosReis, S., Zito, J. M., Safer, D., Soeken, K. L., Mitchell, J. W., & Ellwood, L. C. (2003). Parental perceptions and satisfaction with stimulant medication for attention deficit hyperactivity disorder. *Developmental and Behavioral Pediatrics, 24*(3), 155–160.

Elder, L., & Paul, R. (2001). Critical thinking: Thinking to some purpose. *Journal of Developmental Education, 25*(1).

Kendall, J. (1998). Outlasting disruption: The process of reinvestment in families with ADHD children. *Qualitative Health Research, 8*(6), 839–857.

Klasen, H., & Goodman, R. (2000). Parents and GPs at cross-purposes over hyperactivity: A qualitative study of possible barriers to treatment. *British Journal of General Practice, 50,* 199–202.

Leslie, L. K., Plemmons, D., Monn, A. R., & Palinkas, L. A. (2007). Investigating ADHD treatment trajectories: Listening to families' stories about medication use. *Journal of Developmental & Behavioral Pediatrics, 28*(3), 179–188.

Malacrida, C. (2001). Motherhood, resistance, and attention deficit disorder: Strategies and limits. *Canadian Review of Sociology and Anthropology, 38*(2), 141–165.

Reid, R., Hertzog, M., & Snyder, M. (1996). Educating every teacher, every year: The public schools and parents of children with ADHD. *Seminars in Speech and Language, 17,* 73–90.

Sax, L., & Kautz, K. J. (2003). Who first suggests the diagnosis of attention deficit hyperactivity disorder? *Annals of Family Medicine 1,* 171–174.

Segal, R. (1998). A construction of family occupations: A study of families with children who have attention deficit hyperactivity disorder. *Scandinavian Journal of Occupational Therapy, 65*(5), 286–293.

Wright, S. F. (1997). A little understood solution to a vaguely defined problem: Parental perceptions of Ritalin. *Educational and Child Psychology, 14*(1), 50–58.

Chapter 4

Sala, G., Gorini, A., & Pravettoni, G. (2015). Mathematical problem-solving abilities and chess: An experimental study on young pupils. *Sage Open, 5*(3).

Chapter 5

Barnes, G. R., Cerrito, P. B., & Levi, I. (2003). An examination of the variability of understanding of language used in ADHD behavior rating scales. *Ethical Human Sciences and Services, 5*(3), 195–208.

Kimberlin, C. L., & Winterstein, A. G. (2008). Validity and reliability of measurement instruments used in research. *American Journal of Health-System Pharmacy, 65*(1), 2276–2284.

Menon, A. J. (2014). The strengths and difficulties questionnaire: A pilot study on the reliability and validity of the self-report version to measure the mental health of Zambian adolescents. *Journal of Health Science, 2,* 127–134.

Chapter 6

Cheng, H., Chen Su, C., Yen, A., & Huang, C. (2012). Factors affecting occupational exposure to needlestick and sharp injuries among dentists in Taiwan: A nationwide survey. *PLoS ONE, 7*(4).

Cohen, D., Dillon, F., Gladwin, H., & De La Rosa, M. (2013). American parents' willingness to prescribe psychoactive drugs to children: A test of cultural mediators. *Social Psychiatry Epidemiology, 48,* 1873–1887.

Patel P., Borkowf, C. B., Brooks, J. T., Lasry, A., Lansky, A., & Mermin, J. (2014). Estimating per-act HIV transmission risk: A systematic review. *AIDS, 28*(10), 1509–1519.

Ryan, C. L., & Bauman, K. (2016). Educational attainment in the United States: 2015. In *Population Characteristics, U.S. Department of Commerce, Economics, and Statistics Administration.* Washington, DC: U.S. Census Bureau.

Chapter 7

Barnes, G. R., Cerrito, P. B., & Levi, I. (2003). An examination of the variability of understanding of language used in ADHD behavior rating scales. *Ethical Human Sciences and Services, 5*(3), 195–208.

Yamanka, Fialkowski, Wilkens, Li, Ettiene, Fleming, Power, Deenik, Coleman, Guerrero, Novotny (2016). Quality assurance of data collection in the multi-site community randomized trial and prevalence survey of the children's healthy living program. BMC Research Notes, 1-8. doi: 10.1186/s13104-016-2212-2

Schwartz, Z. (2005). Prescription of medication for ADHD: An unethical practice. *Ivy Journal of Ethics, 4*(2), 18–21.

Chapter 8

Doolan, D. M., & Froelicher, E. S. (2009). Using an existing dataset to answer new research questions: A methodological review. *Research and Theory for Nursing Practice: An International Journal, 23*(3).

Durkheim, E. (1897). *Suicide: A study in sociology.* Mankato, MN: Free Press.

Hughes, S., & Cohen, D. (2011). Can online consumers contribute to drug knowledge? A mixed-methods comparison of consumer-generated and professionally controlled psychotropic medication information on the internet. *Journal of Medical Internet Research, 13*(3), 53.

Paunonen, S. V., & Ashton, M. C. (2013). On the prediction of academic performance on personality traits: A replication study. *Journal of Research in Personality, 47,* 778–781.

Chapter 9

Lester, P. E., Inman, D., & Bishop, L. K. (2014). *Handbook of tests and measurement in education and the social sciences.* Lanham, MD: Rowman & Littlefield.

Chapter 10

Bureau of Labor Statistics. (2015). *Occupational outlook handbook: Registered nurses.* Washington, DC: U.S. Department of Labor. Retrieved from https://www.bls.gov/ooh/healthcare/registered-nurses.htm

Elliott, M., & Lowman, J. (2015). Education, income, and alcohol misuse: A stress process model. *Social Psychiatry and Psychiatric Epidemiology, 50*(1), 19–26.

Jaccard, J., & Becker, M. (2002). *Statistics for the behavioral sciences.* Belmont, CA: Wadsworth.

James, G., Witten, D., Hastie, T., & Tibshirani, R. (2009). *Introduction to statistical learning with applications in R.* New York, NY: Springer.

O'Connor, P. J. (1990). Normative data: Their definition, interpretation, and importance of primary care physicians. *Family Medicine, 22*(4), 307–311.

Piovesana, A., & Senior, G. (2016). How small is big: Sample size and skewness? *Sage Open.*

Chapter 11

Bernard, R. H., & Ryan, G. W. (2010). *Analyzing qualitative data: Systematic approaches.* Thousand Oaks, CA: SAGE.

Burawoy, M. (1998). The extended case method. *Sociological Theory, 16*(1), 5–33.

Covert, J. (2003). Working women in mainstream magazines: A content analysis. *Media Report to Women, 31*(4), 5–14.

Deegan, M. J. (2007). The Chicago school of ethnography. In P. Atkinson, A. Coffey, S. Delamont, J. Lofland, & L. Lofland (Eds.), *Handbook of ethnography* (pp. 11–25). Thousand Oaks, CA: SAGE.

Geertz, C. (1973). *The interpretation of cultures: Selected essays.* New York, NY: Basic Books.

Glaser, B., & Strauss, A. (1967). *The discovery of grounded theory: Strategies for qualitative research.* Chicago, IL: Aldine.

Lipka, M. (2015). Americans' faith in God may be eroding. Retrieved from http://www.pewresearch.org/fact-tank/2015/11/04/americans-faith-in-god-may-be-eroding/

Murphy, C. (2016). Q&A: Why are women generally more religious than men? Retrieved from http://www.pewresearch.org/fact-tank/2016/03/23/qa-why-are-women-generally-more-religious-than-men/

Chapter 12

Creswell, J. W. (2013). *Qualitative inquiry and research design.* Thousand Oaks, CA: SAGE.

Saldana, J. (2013). *Thinking qualitatively.* Thousand Oaks, CA: SAGE.

Strauss, A., & Corbin, J. (2015). *Basics of qualitative research.* Thousand Oaks, CA: SAGE.

Wagman, P., Björklund, A., Hakansson, C., Jacobsson, C., & Falkmer, T. (2010). Perceptions of work-life balance among working population in Sweden. *Qualitative Health Research, 21*(3), 410–418.

Chapter 13

Cloutier, C. (2015). How I write. *Journal of Management Inquiry, 25*(1), 69–84.

Durbin C. (2004). Effective use of tables and figures in abstracts, presentations, and papers. *Respiratory Care, 49*(10).

Lee, J., & Randolph, K. A. (2015). Effects of parental monitoring on aggressive behaviors among youth in the United States and South Korea: A cross national study. *Children and Youth Services Review, 55*, 1–9.

Pajo, B., & Cohen, D. (2013). The problem with ADHD: Researchers' constructions and parents' accounts. *International Journal of Early Childhood, 45*(5),11–33.

Pajo, B., & Stuart, P. (2012). A comparative review of "how to" books for parents of ADHD children and "how to" books for parents of typical children. *Children and Youth Services Review, 34*(4), 826–833.

Roncaglia, S. (2009). *Feeding the city: Work and food culture of the Mumbai dabbawalas.* Cambridge, UK: Open Book.

Rosnow, R. L., & Rosnow, M. (2011). *Writing papers in psychology.* Boston, MA: Cengage.

Sharafkhani, N., Khorsandi, M., Shamsi, M., & Ranjbaran, M. (2014). Low back pain preventive behaviors among nurses based on the health belief model constructs. *Sage Open, 4*(4).

Chapter 14

Blacher, J., Cohen, S. R., & Azad, G. (2014). In the eye of the beholder: Reports of autism symptoms by Anglo and Latino mothers. *Research in Autism Spectrum Disorders, 8*, 1648–1656.

Duarte, N. (2010). *Resonate: Present visual stories that transform audiences.* Hoboken, NJ: Wiley.

Hoogenboon, B. J., & Manske, R. C. (2012). How to write a scientific article. *International Journal of Sports and Physical Therapy, 7*(5), 512–517.

Lucas, S. (2007). *The art of public speaking.* Columbus, OH: McGraw-Hill.

Reynolds, G. (2010). *The naked presenter: Delivering powerful presentations with or without slides.* Indianapolis, IN: New Riders.

NAME INDEX

SUBJECT INDEX

ABOUT THE AUTHOR

Bora Pajo is a social scientist at Mercyhurst University, Department of Applied Sociology and Social Work. Her research focuses on the daily dynamics surrounding children diagnosed with emotional and behavioral problems and their parents. Bora is passionate about data science and machine-learning algorithms and strongly believes that we can only advance our scientific knowledge if we have high-quality data available. She is dedicated to teaching social research methods and statistical analyses in the hopes of transmitting her enthusiasm of conducting scientific research studies.